Stop Working &
Start Thinking

SECOND EDITION

A guide to becoming a scientist

Stop Working &
Start Thinking

SECOND EDITION

A guide to becoming a scientist

Jack Cohen & Graham Medley
with an introduction by Ian Stewart

Taylor & Francis
Taylor & Francis Group

Published by:

Taylor & Francis Group

In US: 270 Madison Avenue
           New York, NY 10016

In UK: 4 Park Square, Milton Park
           Abingdon, OX14 4RN

ISBN: 0–4153–6830–8

Library of Congress Cataloging-in-Publication Data

Cohen, Jack.
   Stop working & start thinking : a guide to becoming a scientist / Jack
Cohen and Graham Medley ; with an introduction by Ian Stewart.
      p. cm.
   Includes bibliographical references and index.
   ISBN 0-415-36830-8 (alk. paper)
   1. Research--Psychological aspects. 2. Research--Methodology. I.
Title: Stop working and start thinking. II. Medley, Graham. III. Title.
   Q180.55.P75C64 2005
   001.4'2--dc22
                              2005011103

Editor: Elizabeth Owen
Editorial Assistant: Chris Dixon
Production Editor: Karin Henderson
Design & page layout: Erika Pennington
Printed by: Biddles Ltd, King's Lynn, UK

Taylor & Francis Group
is the Academic Division of T&F Informa plc.

Visit our web site at http://www.garlandscience.com

Contents

Abbreviations

ANOVA (analysis of variance)

hCG (human chorionic gonadotrophin)

IgG (immunoglobulin G)

RIC (random is clumped)

Preamble

1

Postgraduate students (variously referred to as postgraduates and Ph.D. students throughout this book) get lots of technical support and advice throughout their studentships. You, the postgraduate scientists, are our principal audience. Texts are already available to help you through the process[1]. There is a mass of literature and expertise for all the equipment, chemicals, materials and so on that you will come across during your research. However, we know of no book that discusses use of the most important piece of equipment: your *mind*. It has been said that doing research is very similar to banging your head against a brick wall, and as a postgraduate student you learn how to do it with as little pain as possible[2]. It might appear presumptuous of us to try and tell you how to think (and it is), but we cannot help but believe that it is necessary.

Many graduands – and others at various stages of their careers – have applied to do research with us. JC had a series of questions at interview, different for each applicant but with a common theme. One of the usual questions was 'What do you do for exercise?' and it got a range of answers, from 'country walks' and 'jogging five miles before

---

[1] Examples are Phillips, E.M. and Pugh, D.S. (1994) *How to Get a PhD* 2nd edn, Open University Press; Tietelbaum, H. (1998) *How to Write a Thesis* 4th edn, IDG Books Worldwide; Dunleavy, P. (2004) *Authoring a PhD Thesis: How to Plan, Draft, Write and Finish a Doctoral Dissertation* (Palgrave Study Guides), Palgrave Macmillan. But a search on your favourite Internet book catalogue will produce many, many more.

[2] Ms (now Dr) Angela McLean said this when she was a fellow postgraduate student with GFM.

breakfast' to 'swimming', often 'nothing' or 'not enough!' 'Not *physical* exercise,' JC would explain, 'You're not joining us for an Olympic medal; what *mental* exercise do you do?' It surprised nearly all of them that mental exercise was a required part of the training (but nearly all of them enjoyed *Zen and the Art of Motorcycle Maintenance* and/or *Gödel, Escher, Bach: An Eternal Golden Braid* when/if they got around to reading them).

Development of an individual from an undergraduate to fully-fledged researcher requires a change in mindset; it is not simply a matter of writing a thesis. As far as we are aware, a Ph.D. is the only degree that qualifies you to teach the same degree: there is no higher qualification in science[3]. These two facts are intrinsically related, as the understanding of the process of scientific research is itself enough to enable you to instruct others through the same maze. We offer you some helpful pointers through this path.

Your studentship is also the only time in an academic career (or, perhaps, the last time in your life) that you can legitimately avoid administrative (and teaching) duties. Use it wisely. Although we realise that financial pressures usually make up for the lack of other commitments, you should see your scientific development as your main goal and compromise it at your peril. Much is expected of you: you must produce a thesis, a new mindset, and scientific advancement.

The ideal is that the Ph.D. candidate knows more about his/her subject than anybody else in the world. When the candidate comes in to the *viva voce* examination, the examiners are hoping to help the student expose this. The thesis is the bread and butter of the procedure, and will sit in the library. The oral examination is that occasion when the worthy student is provoked into demonstrating the almost complete mastery of his/her subject, generating new and interesting questions like never before or again. This is what the examiners examine for – their jam, cheese, meat, and occasionally caviar – they certainly don't do it for the pay!

Certainly in the UK, and generally elsewhere, postgraduates are the lifeblood of scientific advance. It is usually Ph.D. students who are given the 'riskier' projects (i.e. those that might not produce publications), and may find themselves more often at the boundaries of human knowledge. Even if you feel that your supervisor and the original project are keeping you from these boundaries, you should be striving for them. Imagine a supervisor whose last three students studied, respectively, the effects of A on X, B on X and C on X. Guess what project is lined up for you! But there will be a frontier near there, too, if you can find and expose it.

---

[3] A doctor of science (D.Sc.) degree is usually awarded later as a recognition of achievement in science: to prove that the Ph.D. degree was correctly awarded.

There is, of course, a potential conflict here: initially, a supervisor tends to see project and student as two separate entities. But a successful student is one who merges these two into a single entity. Initially, the student will see the supervisor and project as inextricably linked. But a successful student is one who disentangles these two into separate entities.

Only if you think that the project can produce research that will change people's minds should you start it. Ask yourself, 'Where is the *Nature* paper in this?' One of our aims is to help you answer this question. Most of our chapters address ways of thinking about research, about experimental design, for example, and being sensible about statistics.

Chapter 11 looks at what happens when things go wrong, and how you might deal with a variety of problems.

Chapter 12 is our attempt to cover your starting research, maturing as a postgraduate researcher, and presenting your work as a talk and as a thesis. We also advise you about different styles of research group. If you're debating whether to commit yourself, perhaps you should start there. But read the postamble (Chapter 13) too.

We, being biologists, have written this book for biologists, but we hope it may be more generally useful for scientists, and, possibly, social scientists. We apologise here for the necessary biological jargon we use in our examples, particularly names of organisms; you other scientists aren't going to be examined on these, so you can forget them as soon as you've understood our point. The biologists ought to know the names anyway, and a (smaller and smaller, we fear) fraction of them will actually have seen living examples. A more polemic view of this book is that we are attempting to wrest the influence on the design of experiments away from statisticians and back to biologists. We continue to be amazed at the lack of statistical understanding demonstrated by biology postgraduates, and indeed biologists generally. Statistics (like protein chemistry or radiation physics; X-ray crystallography or knowledge of real ecosystems: 'natural history') brings many tools to bear on biological problems. This lack of understanding, we feel, is due to the lack of explanation of the role of statistics within biological experiments. We hope to correct this, so that you can use statistics more constructively, not only after the results are in but also before you actually do anything, while there is thinking time. (Yes, we know this turns our title around; we don't mind this, nor will you.)

Introduction 2

If you're going to stop working and start thinking, you need to be aware of the pitfalls that await those who leap to 'obvious' conclusions – often without being aware that they have done so. Thinking is a skill that must be learned, not something you can do off the top of your head, so to speak.

2.1 Please bear with us

A cameraman working for a television company that films wildlife was out in the wilderness somewhere when he spotted a bear 1 mile to his east. Unfortunately the sun was in a bad position for filming, so he turned north and walked straight ahead for 1 mile. Then he saw the bear – which had not moved from its initial position – 1 mile to the south, and using a zoom lens he had no trouble filming it. So: what colour was the bear?

(All distances and directions are exact, and the Earth may be assumed a perfect sphere. If there is more than one possibility, specify them all.

Ignore minor colour variations: 'reddish-brown with a few dark patches' is the same as 'brown'.)

This little puzzle will be familiar to most of you – though probably in its less politically correct version in which the man shoots the bear. And you are probably wondering what it is doing in a book on experimental methods. Fear not: all will shortly be revealed. But before you dismiss the answer as obvious, have a go at the interplanetary version. Same scenario and conditions, but this time the cameraman is on the distant planet of Argyris III, where blue bears are common all over the north polar cap of frozen methane, the bears on the south polar cap of frozen carbon dioxide are purple, and there are varieties of green, yellow, and pink bears around the equatorial regions. Oh, and Argyris III is, by a remarkable coincidence, exactly the same size as the Earth.

If you got 'blue', think again.

The point of this puzzle is to convince you that even in a subject as clear-cut as geometry, it is easy to make unwarranted assumptions. In the far more messy world of experimental science, it is even easier to make unwarranted assumptions. It might seem that assumptions are irrelevant there – what matters are data, and there are no assumptions in observational data. However, there are. Even if the data themselves are completely 'objective' – which is typically not the case – they must be interpreted in order to convey useful information. This interpretation step involves many assumptions by the experimentalist, often tacit, perhaps not recognised at all. Less obviously, the choice of what things to measure, and how, also involves assumptions. Unless you make yourself aware of what these assumptions are, your experiment may give misleading results. So it pays to be careful, and to think the matter through before you wade into a programme of experiments.

The rest of this book will show you what can go wrong if you don't, and takes a closer look at the causes of error and ways to avoid them. The most important is to keep your brain in gear. Mimicking other people's procedures might seem to be a safe way to design an experiment – after all, the referees let them publish their work, so it must be OK. Sadly not – well, not always – as we will see.

Back to the bear. The answer to the puzzle in its traditional form is 'white'. At first sight the information given has no bearing (pun unintentional) on the animal's colour, but actually it does. In most parts of the world, if you start at the bear and go 1 mile west (to where the cameraman was), then 1 mile north (where he went), then 1 mile south (where he saw the bear), you do not get back to the bear (*Figure 2.1*) If we can work out where the cameraman was, we may get a

Figure 2.1

On a flat world, or away from the poles, the geometry doesn't work.

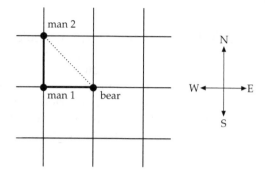

useful angle on the bear's species, hence colour. And a few moments' thought shows that if the cameraman's final position is at the north pole, then the geometry works out OK if the bear and the initial position of the cameraman are suitably chosen (*Figure 2.2*). So it was a polar bear, and therefore white.

Sure, maybe a brown bear had escaped from a circus that was being transported over the pole by air and the plane had crashed. In puzzles, that kind of answer is generally agreed to be cheating – language is too imprecise to rule out every conceivable get-out of this kind; you have to enter into the spirit of the enterprise. In experimental science, you have to worry rather more about freak effects that may have propelled your experiment along an unusual trajectory.

Figure 2.2

The geometry of longitude and latitude at the north pole is different.

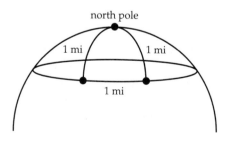

What about the multicoloured bears of Argyris III? They are actually silicon-based ursinoids, but 'bears' will suffice. The same geometry works at the north pole of that alien world, so the bear is blue, yes?

No. It might be blue, but it could be purple. Indeed, if the Earth's Antarctic zones sported purple bears, then that would be a possible answer on this planet too. Fortunately for puzzle-setters, Antarctica is a bear-free zone – barring crashed circuses.

At just the right latitude, roughly a sixth of a mile from the south pole, there is a line of latitude that forms a circle exactly 1 mile in circumference. If the bear is anywhere on that circle, and the cameraman is in the same position as the bear, the geometry also works out OK (*Figure 2.3*). Agreed, he wouldn't see the bear if he tried to look right round the circle, but we said 'spotted a bear 1 mile to his east'. He has no trouble spotting the bear, since he is right on top of it, and it is indeed 1 mile to his east! So on Argyris III it could also have been a purple bear.

Figure 2.3

The geometry is different at the south pole, too.

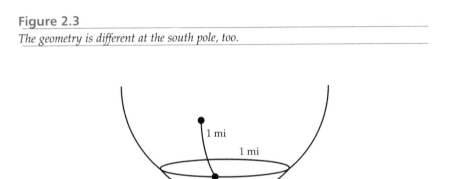

On an Earth-sized spherical planet, no such geometrical peculiarities occur anywhere near the equator, so the green, yellow, and pink bears are out. The correct answer, then, is 'either blue or purple'.

Have we listed all the possible places for the bear to be? Not at all. There is also, rather close to the pole, a circle of latitude exactly half a mile long. Again the bear is 1 mile to the cameraman's east if they are both in the same place – you just have to go twice round the circle. Is that all? Of course not: the same kind of thing happens for circles of latitude of length 1/3 mile, 1/4 mile – indeed, $1/n$ mile for any whole number $n < 1$. Just go n times round to get your mile.

None of this affects our answer, though, for all bears in such places are again purple. On the planet Bearblebrox, mind you, where each line of latitude is the habitat of its own unique colour of bear...

So that's the lot, then? Think about it!

2.2 Take your time answering this one...

Which would you prefer: a pay rise of £500 every 6 months, or a pay rise of £1600 every year?

Clearly, the latter. The first gives you a salary increase of £1000 every year, which isn't as good.

Happy? Let's think it through on an example. Suppose you start out on 1 January of the year 2000 earning £10,000 (the actual figure makes no difference to the argument, it turns out). Assume the pay rises occur on 1 January and 1 July for the six-monthly increase, and on 1 January for the annual one. Let's tabulate your earnings, year by year, in the two cases:

Year	£500/6 mo	£1600/year
Jan 2000	5000	5000
Jun 2000	5500	5000
Jan 2001	6000	5800
Jun 2001	6500	5800
Jan 2002	7000	6600
Jun 2002	7500	6600

Note that the earnings in the last column have been divided by 2 to get the figure received in each 6-monthly period following the date.

Fascinating, isn't it?

2.3 A tale of two goats

This elegant probability puzzle is about a game show. A simple mathematical model of the show goes like this. The contestant is shown three closed doors. Behind one is an expensive car, and behind the other two are goats. The contestant wins whatever is behind the chosen door. The twist is that after the contestant has made the choice, the host – who knows where the car is – points to another door, different from the chosen one, opens it to reveal a goat, and asks the contestant if they wish to change their choice of door.

Let's agree some ground rules. You may assume that the probability of the car being behind any given door is one in three. The host knows exactly where the car is, so he can always choose a door that reveals a goat, whatever the contestant chooses.

Here's the question. Does the contestant's chance of winning the car improve if they now change their mind and switch to the other unopened door?

Idea 1: The host's actions leave the contestant faced with two doors. They have no idea which hides the car. It is just as likely to be behind the door that they first chose as it is to be behind the other door. The probability of winning is $1/2$ whichever door they choose, so they might as well stick to their first choice.

Idea 2: The probability of winning the car improves from $1/3$ to $2/3$ if the contestant swaps. Here's why. If their first choice was right (probability $1/3$), they will lose the car if they swap. But if it was wrong (probability $2/3$), they will win the car if they swap. So two-thirds of the time swapping should secure the car.

Idea 1 sounds pretty convincing. But bear in mind that by the same argument, they might as well swap, and if idea 2 is right, they definitely should. Curiously, hardly anybody who believes idea 1 wants to swap. This is very strange, and suggests that psychology is at work as well as probability. Leaving such considerations aside, we seem to be faced with two plausible mathematical arguments which flatly contradict each other. The basis of idea 2 is simple: the probability that the contestant's first choice will secure the car is $1/3$. It seems difficult to dispute that, and all else follows logically.

On the other hand, idea 1 has an equally plausible basis: since the contestant knows that there's a goat behind at least one of the doors they failed to select, how can it help to be told which? Two doors on offer, each equally likely...

Which – if either – is right?

Probabilities are numbers that tell us how likely certain events are. An event with probability $1/3$ will occur about one time out of three, in the long run. This interpretation immediately rules out one possible justification of idea 1. Perhaps the probability of the car being behind the door that the contestant originally chose was $1/3$, but it somehow becomes $1/2$ once the host opens another door. No. Imagine running a long series of trials – say, three million. Roughly one time out of three – on approximately one million occasions – that's the right door. And if the car was behind it before the host revealed a goat, it will still be behind it afterwards. If the choice was wrong – approximately two million occasions – it stays wrong. So it will be the right choice on exactly the same occasions as before – approximately one million of them. That's one time in three: probability $1/3$. Which choice is right or wrong doesn't change, so the probability of success can't change either.

Presumably one of the two approaches conceals a tacit assumption that turns out to be false if you think about the problem correctly. Which? Faced with this kind of dilemma, it often helps to get some independent evidence. It's not hard to write a computer program to simulate the problem, run it a large number of times, and count how often the contestant succeeds and how often they fail. Do this with a no-swap strategy, do it again with a swap strategy, and compare. I once ran 100,000 trials. The no-swap strategy gave the correct answer in 33,498 trials, and the wrong answer in 66,502. With the swap strategy the numbers were exactly the other way round (not a coincidence but a logical necessity: one strategy is right precisely when the other is wrong – the main observation behind idea 2). The corresponding probabilities of 0.33498 and 0.66502 are convincingly close to $1/3$ and $2/3$.

If idea 1 was correct, we would have expected numbers very close to 500,000. We got no such result. So the experimental evidence suggests that idea 2, the counter-intuitive one, is right, and that the intuitively plausible idea 1 is wrong.

Here is an argument that in effect provides a mathematical proof of the validity of idea 2. Number the doors 1, 2, and 3 so that we can list possibilities. Suppose, for the sake of argument, that the contestant chose door 1. (The same kind of thing happens if they chose doors 2 or 3.) There are three possibilities, and each is equally likely (because the probability of the car being behind any particular door is $1/3$). An italic goat indicates a door that the host can safely open.

No-swap strategy

door 1	door 2		door 3
car	*goat*	*goat*	win
goat	car	*goat*	lose
goat	*goat*	car	lose

Swap strategy

door 1	door 2		door 3
car	*goat*	*goat*	lose
goat	**car**	*goat*	**win**
goat	*goat*	**car**	**win**

The no-swap strategy wins one time out of three; the swap strategy wins two times out of three.

What, then, is wrong with idea 1? The assumption that with only two doors to consider, the car is equally likely to be behind either of them is correct if the host opens a goat door before the contestant makes their choice. However, it is wrong if the host's choice depends upon what the contestant has already chosen, and that's the case here.

A few years ago I watched a television programme about a man who was winning substantial sums of money by betting on the weather. He would make very complicated bets, along the lines of the number of rainy days in April minus the average temperature (Celsius) at midnight in February will be an even number. A meteorologist was interviewed at one point in the programme, and said roughly this: 'On any given day, either it rains or it doesn't. So the chance of rain is 50-50.' Oops. With his brain in gear on another occasion, he would have realised that this is far from true. The probability of rain on any given day is quite low in the summer, much higher in spring and winter. So it can't be 50-50 all the time. For the UK – and most other places on the globe – meteorologists have compiled lengthy records of rainfall: it would be easy to provide specific figures here. Given that a meteorologist caught off-duty displays such a poor grasp of the basic probabilities of his own professional area, pity the poor bookmaker who accepts complex bets on the weather. Now you can see why the punter had a habit of winning.

What was the mistake? It was to assume that when one is faced with two choices, they must be equally likely. The same mistake is what makes idea 1 so attractive.

This fallacy can (for some people) be driven home by a variant on the game show which uses cards. Take a standard pack of 52 cards: deem the ace of spades to be the car, and the other 51 cards to be the goats. The host offers you the pack and lets you choose a card. You win whatever corresponds to that card. It is face down.

What is the probability your card is the ace of spades? 1/52. Ah, but now the host turns over the remaining cards so that he can see them but you can't, and starts throwing them face up on the table, with no ace of spades appearing. He continues to do this until he holds a single card, and still no ace of spades lies on the table. He places that sole remaining card next to yours, face down, and invites you to change your choice to his card.

Two cards – 50-50? Not at all. Your card will be right one time in 52. His card will be right 51 times out of 52.

How come? Before he starts discarding on to the table, you know that at least 50 out of his 51 cards are not the ace of spades. But if he does hold the ace of spades, you don't know which card it is, and choosing one at random from his 51 cards does not improve your chances. After he discards 50 cards, though, you know exactly which one of those 51 to choose. If your card was the ace of spades all along, then swapping won't do you any good; but if the ace of spades was among the host's 51, you know exactly where it is now. On 51 occasions out of 52, in the long run, swapping will secure the ace of spades. On one occasion out of 52, in the long run, not swapping will secure the ace of spades.

No contest.

Those two cards look exactly the same, lying there face down. But the procedures that led to those cards are very different. Yours was selected by a random choice out of 52, with no knowledge about which was the ace of spades. The host's card was selected by non-random choice out of 51, with complete knowledge about which was the ace of spades. The apparent symmetry is deceptive. Now, with luck and a following wind, you will see that the same goes for the game show. The host's choice does give you new information, and by exploiting that, you can improve your chances.

If you're not convinced, try it with a pack of cards and a friend. To be really convinced, play for money.

2.4 Get your brain in gear

How do these three puzzles relate to the central topic of this book – how to be an effective (experimental) scientist?

The main message is straightforward. Even in areas where the rules are clear, the question is well posed, the data are complete and accurate, and there is an obvious solution, the human mind often gets the answer wrong. This happens because humans have a tendency to leap to conclusions. This isn't always a bad thing; indeed, evolution probably equipped us with the ability to make intuitive leaps because a quick-and-dirty decision may have greater survival value than a more accurate one that takes longer. When we are building up science, however, immediate survival value is not as important as long-term durability. The big problem with these intuitive leaps is that we are seldom consciously aware of them. If we don't know they're present in our thinking, we can't take steps to check whether they work. 'It's not what you don't know: it's what you know that ain't so.'

There are some other messages, too. The bear puzzle shows that vital information may at first sight appear irrelevant – all that fuss about distances and directions, and then you get asked about the bear's colour. The puzzle also shows that you need to distinguish between 'normal' behaviour (the geometry of the Earth's latitude/longitude coordinate system in the habitable regions, where a horizontal/vertical coordinate grid on graph paper is a good approximation) and 'singular' behaviour (near the poles, lines of latitude form tiny circles, and all lines of longitude cross at the pole). This effect, incidentally, results from our choice of coordinate system – the shape of a sphere near its poles is exactly the same as its shape near any other point. In a sense, the whole puzzle is an artefact of our choice of measurement system – and that can happen in science, too. The final message is especially important: having figured out one solution, it is all too easy to assume that it's the only possible one. In science, having puzzled out one plausible explanation of some phenomenon, you must step back and ask yourself whether there are alternatives (which might be more subtle, too).

The pay rise puzzle tells us, again, how readily we leap to obvious conclusions, and that the obvious may be false. And it makes the point that sometimes this kind of error can be avoided by thinking your way systematically through a simple example.

As for the goats – much the same message, but more powerfully delivered. Many people refuse to believe that changing your mind actually does improve the chance of winning. They are so attracted to the short cut (the two remaining doors are equally likely) that no amount of evidence can get the conviction out of their heads[1]. No enumeration of possibilities, no computer experiments, will ever convince them that they are wrong. Even those of us who are convinced still find difficulty wrapping our heads around the reasoning – there's often a faint feeling at the back of our minds that we've failed to grasp something crucial. For the untrained human mind, probability is not an intuitive concept[2]. This is one reason why we make bad decisions about gambling, why many of us find it impossible to trust statistics, and why we have such a poor understanding of the risks inherent in various activities.

There is a serious warning here, because statistical analysis forms the core of traditional experimental design. Learning statistical methods by rote is very dangerous: unless you have some idea where they come from, what they're for, and what their limitations are, you can make a major mistake and have no idea that you've done so. The journals contain many published, refereed papers that misinterpret statistics. Entire areas of research have sometimes grown from these initial misunderstandings, leading to scientific 'urban myths' – theories and experiments that have no factual basis – that it takes decades to dispel.

Even if you have a very strong 'gut feeling' that some assumption is obvious, you may be wrong. Examine your assumptions – especially the subconscious ones, the ones you don't even realise you are making.

---

[1] If the host of the show can decide whether or not you get a chance to change your mind, the answer depends on how he does this. Suppose, for example, he offers you that option only when the door you chose hides the car. A subconscious worry about this kind of effect may be part of the reason why some people can't agree that swapping is better.

[2] See our Bayesian examples (Chapter 8, Section 8.2).

Science like what it is done

3

3.1 What is science?

Science is what scientists do, what children learn in the science class at school, and what the science textbooks contain. It is also what the media science correspondents tell us about, and it is supposed to centre on what goes on in laboratories. These are all different activities, and none are easily defined, or even described. Yet they are all called 'science'. This is a problem with definitions, not with science. To illustrate this, consider another concept which is difficult to define. Think about what we mean by 'alive'. You're alive, we're alive (though perhaps not by the time you read this), an oak tree is alive, an amoeba is alive, and so is a bacterium. But a virus? A crystallised virus? A mitochondrion in a cell? A 'dead' leaf drifting down off a tree in autumn (it might have a couple of cells in it which, by a heroic effort, we could persuade to grow into a new tree)? One of your red blood cells, without a nucleus? Each of these, for some purposes, might usefully be considered to be alive – or not. Equally, for science: the teacher showing the infant class woodlice; the museum curator inventing an explanation of steam pressure for the label on an exhibit; the mathematical physicist worrying at an equation to make sense of some radiation data; an ecologist sorting through a pile of moths caught by a lamp; a university professor lecturing about transition elements; her postgraduates arguing about their experimental results over coffee (naughtily[1]) in the lab – these are all science, or not, according to your concern of the moment. Don't worry about the definitions.

Science is best thought of as a great *variety* of activities, deriving from a few kinds of enquiry in the Western world over the last 300 years or so. It is, by many criteria, the most successful way of understanding the world that we have invented. To some extent this is because 'understanding', in Western culture, has come to *mean* accepting science-based explanation in the language of science. But it is agreed that our command of technology, of modern medicine, and of computing and communicating have all derived from that under-standing. No other style of thinking has been able to achieve this. At every stage, too, each of these technical subjects has provided fuel for the advancement of its own kind of science. We do not ask, 'Which came first, the technology or the theoretical background?' Each clearly feeds the other, forming what we have called a 'complicity'[2] between theory and practice. These different activities, feeding each other, have formed a progressively larger part of the accumulating capital of human culture, what we have called our 'extelligence'[3], complementing our intelligence and fuelling each intelligence as it matures and learns.

Our favourite description of science is by Ian Stewart: 'Science is our best defence against believing what we want to.'

There is a strange property of scientific learning: it destroys itself. Successive generations of scientists can be cleverer, but needing to know less, remember less. This is because scientific theories make many facts irrelevant. 'Theories destroy facts,' said Medawar. Here is a silly, but useful, way of explaining this: before Newton invented gravity (or did he discover it?), people had to remember that flames pointed up, that apples and people and pebbles and porcelain cups fell down, that the Moon's orbit was such-and-so. Afterward, all we need is 'Every object in the universe attracts every other object in the universe with a force which is proportional to the mass and inverse-ly proportional to the square of the distance', and we can then *work out* that water doesn't flow uphill! And we can work out the orbit of Mars. And all the other applications of the law of gravity become trivial, not worth learning by heart because you can get to them by remembering the law and applying that. It happens all over science. Classifying plants enables us to talk of dandelions, instead of saying, 'There's a plant with a yellow flower ... and there's another one ... and another.' And when we have the angiosperm family *Compositae* we can put dandelions, daisies, different daisies, dahlias and chrysanthemums into the same thought package: they all have flowers made of lots of little flowers. And once the chemists have

---

[1] This is a capital offence according to safety officers.

[2] Cohen, J. and Stewart, I. (1994) *The Collapse of Chaos; Simple Laws in a Complex World*. Penguin, Viking, New York.

[3] Stewart, I. and Cohen, J. (1997) *Figments of Reality*. Cambridge University Press, Cambridge.

named the elements, and got words and rules for combining them, or for analysing compounds, we can handle the combinations like a kind of molecular Lego™, much simpler than the alchemists' formulae, *and* more reliable. Sodium chloride, uranium sulphate and mercapto-ethanol are all simply labelled parts of the chemists' network of rules. Other human concerns, like music or the other arts, do not destroy their history like that; it makes science much easier to learn.

In order for science to grow like this, to move what was complicated and difficult to understand into a simple part of the school curriculum, we must constantly check that what we're discovering – or inventing – is 'good' theory. It's no good just guessing and putting this into the books (though lots of people *have* done just that, as you may have discovered). Science has several agreed ways of validating, or rejecting, new ideas. Scientists are only rarely *taught* what these are – mostly we acquire science, like learning to swim by practising swimming in a pool with many good swimmers. This book attempts to put some of these practices, different swimming strokes and how you can't easily breathe under water, down on paper. We know that we can't teach you to swim from a book, but we still think that it's a good idea to think about what you're doing, though we do know that most scientists just dive straight in and begin working. And that even after the publication of this book, they will continue to do so.

3.2 The methods of science

Science asks questions, and it has a small variety of ways to look for answers. These are *observation, measurement, investigation* and *experiment*. Different kinds of problems need different approaches for their solution, and one of the ways the experienced scientist knows which to use is that she's got it wrong many times in the past! This cannot be said too often, emphasised too much. *Ignorance,* recognised, is the most valuable starting place; all scientists should have many stories about where they were sure – and wrong; where they were ignorant but didn't know it. The big problem for all of us, indeed, is that it feels just the same when we're sure and wrong as when we're sure and right!

That is how the intellectual construction of science proceeds: apparently sturdy intellectual scaffolding turns out to be useless, or broken, or much better put in the other way round. If you do a laboratory exercise, perhaps when you are a student, and everything works as it 'should', according to what it says on the instruction sheet, *you have learnt nothing* (except perhaps a bit of cookery). If, however, you've had to modify the apparatus, change the temperature, use a different indicator, rewrite some of the code, or double the concentration before it 'works', then you have investigated the possibilities around the experiment you've done. You may actually begin to *understand* it. You may be able to explain it to somebody else and improve *their* understanding.

That is why we must learn to observe, to see that we always choose among observations; and why, once we begin to understand, the first thing to become obvious is that we should have made *different* observations. That is why we must learn to measure, and to assess how repeatable, reliable and inaccurate our measurements are. We must learn to investigate, to recognise the kinds of questions-of-nature that will provide answers, to guess when pilot experiments should be succeeded by more expensive, usually more complicated designs. Lastly, we scientists must all discover our own ways of modifying the universe to make it perform revealing tricks (if we are to be more than journeymen). We call this experimenting, and we must learn to experiment.

Many scientists have never been taught how to experiment – but they can do it very well. It is important to remember that you don't need to understand in order to do: people made babies long before anyone knew about eggs and sperm; babies learn to see without having any clue about optics. In the same kind of way, many people do good science without understanding how it works. Those scientists who do understand may have learned some of it from books, but most learned from working with other scientists, and especially by the willingness of colleagues to be usefully critical. Supervisors of Ph.D. students are supposed to instruct their pre-scientific postgraduate students with the elements of scientific procedure, but most of them have become successful scientists without ever having made their methods explicit. It is as if swimming was learned by swimming beside better swimmers, who themselves learnt in the same way. As we said above, it would be difficult to put this kind of learning and teaching in a book.

We believe that there *are* ways of making these scientific techniques explicit, but not as a kind of exercise in the theory of science. We think that scientific methods make best sense when applied to real cases, in the context of real puzzles. Therefore, much of this book is anecdote and experience. Most of the examples we use here will be biological, because most of our experience of experimental science is in this area – but we will put the occasional physical, chemical or psychological riddle in the text, too.

There is an issue here that must be addressed, and our position on it is contrary to present trends in education, perhaps especially in science education. We believe that scientists, and especially aspiring scientists embedded in their education, should be doing practical things with real-world problems. Not exercises remote from the real world (but ones to which the teacher/professor *knows the answer*, because he's given this same exercise to students for 30 years!), nor model systems on computers (their apparent educational advantage – that *they always work* – is precisely what we think makes them pretty useless). We like students to gather/measure/breed their own organisms, get their hands dirty working with them, do their own washing-up, write/modify their own programmes. We deplore the existence of kits for the education of students, but agree that it's usually better than not doing the exercise at all but only seeing pictures on a computer screen. JC's colleagues in the 1980s discouraged the breeding of Siamese fighting fish (*Betta splendens*) for practical classes, or using living zebra-fish eggs to draw the blood system, but encouraged the use of JC's films from previous years – much cheaper, worked every time (but JC thought: pretty, but not educative). When they proposed substituting film of previous student groups doing fieldwork (collecting and measuring organisms on seashores) for students actually doing it, JC left the institution (after 30 years).

There is another prejudice we will confess here. Some students are lucky; some are unlucky. We don't think this is in our stars or our genetics, but in the way we have been taught (or taught ourselves) to deal with life's problems/opportunities/challenges. The luckiest postgraduate students we have supervised have been those who foresaw what *could* happen as they mentally reviewed their programme of the next couple of days, and took care of those unexpected events which could bite them in the ankle[4]. The unluckiest students were those who copied what someone else did, or did exactly as they were told, without putting their mind in gear beforehand and running it through mentally to see what might go wrong. We really don't like it a bit when students do exactly what they're told!

Those wonderful kits that have all you need to do tissue culture or DNA comparisons, or the microscopes that automatically give you the right phase ring in the substage for the objective you choose, or the blood-analysis machines that give you a read-out with 30 variables measured, are like buses: they take you where *they're* programmed to go. In this book, to extend the metaphor, we want to teach you to drive a car, so that you can determine the objects and the goals of the journeys you make.

---

[4] Dan Dennett's book *Freedom Evolves* (Viking 2003) says that the only free will worth having is to make the inevitable evitable: Odysseus took care of the Sirens' attraction, prospectively, with some rope and some wax, you will recall.

3.3 Where do scientists get the questions they ask?

This is itself a very difficult question, as we'll discover in later chapters. There are, however, two straightforward answers, both of which you have all experienced. But these two answers don't help with what we think are the important, deep questions; how we question the way the universe works. Too many people, though, *do* believe that they really *are* how scientists find their questions: they get them from cleverer scientists, and never mind the universe!

Here is the first, trivial source of questions. You go into an exam room, sit down, turn over the exam paper and read the questions. They might say, 'What is the protein in blood which carries oxygen from the lungs to the tissues?' or 'How does what you know of the structure of haemoglobin account for the Bohr shift in its oxygen-carrying capacity?' They might ask, 'Why is the sky blue?' or 'What are Schrödinger's contributions to the theory of wave mechanics?'

These questions are the wrong way round, for us in this book. They are the authority figure directing your attention to areas where *he* wants *you* to think. They are somebody else trying to find out what *you* know, hopefully by giving you a chance to air your knowledge. They are testing your knowledge base, which must be there for you to ask useful questions, but which probably doesn't help you to design those questions.

The other straightforward answer to where we get the questions is the remote authority one. You are doing a task, like centrifuging cells from a suspension or baking a cake; 'How do I calculate g forces from the diameter of the centrifuge head and the speed?' is the same kind of question as 'What mark do I set the oven on to get 210 °C?' Questions like 'What is the best temperature for baking this kind of cake?' or 'What is the best probe to use for this DNA sequence?' are of the same kind. Not very different are 'How many spiders are in this field?' and 'What is the DNA sequence of the gene whose mutation causes diabetes?' The remote authority, in the *Textbook of Field Studies* (we invented this), says, 'To find out how many spiders, you need to …'. The remote authority for the DNA sequencing is almost certainly a website.

We don't want to trivialise this type of question. Scientists, like all of us, have to answer *this* kind of question all the time, using scientific cookbooks or the expertise of our colleagues. We return to this in Chapter 5; all scientists need to be competent, thinking measurers and observers. However, some of our colleagues pretend that all of the questions that they, and all of serious science, are concerned with are of this kind. What? and how?, they claim, are the whole business of science. We think this is 'nonscience'[5] (to rhyme with 'conscience' –

---

[5] Ford, B. (1971) *Nonscience*. Wolfe Publishers, London.

and, nearly, with 'nonsense'). Scientists – and there are many – who believe that science is 'only' about this kind of authoritative 'truth', about what? and how? and never, never about why? set different kinds of questions in final examinations in university. They talk of model answers to the questions they set, unique answers that would earn 100%, when all other answers would earn less. They want *their* version of the truth forever from their students. We and other scientists of like mind don't believe that there is only one such right answer to most scientific questions. Even in the difficult questions we set in your final examination papers for your first degree in science, there should be *many* more ways of earning 100% than of earning 30%.

Think about this; it sits at the base of what your culture thinks about truth (we discuss cultural differences in our approaches to 'truth' in Chapter 12). Some of you are one-answer people – who are happier with the idea that there is a right answer, but that we probably haven't got it yet; science is converging on it[6]. Some are not happy unless they have several answers, preferably with a few contradictions, and believe that science requires divergent explanations. You can be a good scientist with either a convergent or a divergent approach.

Both are necessary for science. Convergent scientists set up the framework of the subject – they are the umpires for the current set of rules. They decide what things are called, where and what the fundamentals of the subject are, and what the undergraduate syllabuses contain. They frequently edit journals and are on grants committees. Divergent scientists challenge the rules continuously, and provide the impetus for changing them. They often cannot say what a particular grant proposal will result in. The best scientists can wear either hat.

Whichever stance you have, it feels exactly the same to be sure and wrong as it does to be sure and right. You will learn not to judge your correctness on the basis of how sure you are. You should learn that the fact that others are convinced they are right has no evidential weight[7]. Much better to believe people who say, 'I don't know' or 'I might be wrong', since they might well be. Just like a dancer, the best scientists are always well balanced.

Science, we think, is about important questions, which could have several very different, mind-changing answers. In systematics, for example, authoritative questions have only one, 'right' answer, and the convergent scientist works well here, classifying and naming animals and plants. For a reproductive biologist, the sight of two porcupines is a question; for an astronomer, the sight of the sun's rays apparently forming a fan in the distance, diverging from each other instead of being parallel, is a question (*Figure 3.1*).

---

[6] A characteristic remark is 'Make up your mind – which is it?'

[7] *The Devil's Dictionary* by Ambrose Bierce defines 'positive' as 'mistaken at the top of one's voice'.

Figure 3.1

Porcupines do it against the prickles, and sunrays do it divergently.

For JC's Ph.D. thesis, the existence of birds and mammals with some white and some coloured patches was a question (see *Box 12.1*). The answers may simply fill a gap in our knowledge, as in 'How *do* porcupines manage to copulate?' or 'What is the colour of mercuric oxide?' Or they may resolve apparent paradoxes, like the sunrays being parallel but appearing to converge and diverge. These answers, though, are but a short step behind why? questions, like 'Why do porcupines make it so difficult?' or 'Why do many oxides come in several different colours?' Designing the kinds of observation and experiment to answer these different kinds of questions can be easy, difficult or, sometimes, even in principle, impossible, as we'll see. But they do not result in one, right, answer, as the trivial questions do. That is why convergent scientists discourage their colleagues from asking them; very often you can't 'make up your mind' which is the right answer.

3.4 Prediction and predilection

We test our understandings of the nature of things and processes, or indeed of nature, by making predictions which arise from our explanations, and testing them. The philosopher Popper's great insight was to see that however many times we get our predictions 'right', the explanation is not confirmed[8]. One wrong prediction, however, can disqualify an explanation. Therefore, effective use of a scientist's time, in weighing up different explanations, should require her to search for easily disproved predictions. We should not use predilections (tendencies) of the situation, though that is in fact very commonly done. We should seek 'fragile' predictions, especially if we are testing important, mind-changing new explanations. A theory which predicts the first rain in Manchester/Seattle next year on 30 March is greatly to be preferred to one which predicts that it will rain in Manchester/Seattle next year (Manchester and Seattle have a predilection for rain – and vice versa!).

Most scientists do, however, test what they are doing by 'obvious' expectations, predilections of the system. They *like* an experiment whose result is comfortably predictable, confirming their prejudices and satisfying the promises they made in the grant application which is funding the work. Predicting that a new protease will be related to one of the known protease families, or that elementary particle collisions at higher energies or new angles will generate such-and-so patterns, is like predicting rain on the west coast, or sunshine in Fiji. Theories are not tested; prejudices are confirmed.

This kind of experiment is comforting. Experiments take time and money to perform – you need experiments that 'worked' to put in

---

[8] Popper, K.R. (1963) *Conjectures and Refutations*. Routledge and Kegan Paul, London.

your thesis and laboratory reports, which seem to justify your pay packet and existence. The administrator of any laboratory much prefers you to do experiments that work. But, of course, experiments that advance science are the ones that all your colleagues tell you *won't* work. And the important thing is that they usually don't, because your colleagues are usually right. Paradoxically, *we* think that you should be encouraged to do experiments that, prospectively, and from an orthodox position, look as if they won't 'work'.

Generally, as we will see, it is appropriate to do some 'safe' experiments during the normal course of scientific work. It is very important to be able to do the routine laboratory exercises with the standard apparatus. These can be called protocol experiments[9]. Just as the lorries on UK roads are using the label 'logistics' more and more, so experimental and clinical kits have 'protocols': instructions on how you're supposed to do them. It is when such working needs to be supplemented – or replaced – by *thinking* that crucial tests are needed, varying or adding to the protocol. Usually, on first encountering a problem there is an obvious thing to do – something which fits in with all the other work and theories – the sort of experiment that will confirm predilections by making safe predictions. These experiments at least confirm that you know what you are doing. Occasionally, however, somebody does what everybody calls clever science. They make a prediction that is unexpected – the result will – or would – shock everyone if it comes off. Nearly always it does *not* come off. The nature of these more exciting predictions is that they disprove an apparently 'fundamental' theory, and this, in consequence, means that we have to change a lot of our other theories about the way things work.

It would be nice to believe that *all* scientists are doing some of these dangerous, unpredictable experiments, and that they never get known because they mostly don't work. What we fear is that very few people *do* do them, because they cannot think of anything like that, or they lack the courage to try. One of the major problems is that, because science is communicated in stories, it is so difficult to *disbelieve* – it all hangs together so nicely. After a while (and from birth to age 21 years *is* quite a while, especially if you *are* 21), you become hooked into the mindset that there is nothing left to discover, just a few more things to describe. We hope that reading this book will give you a clue as to how to look at science in a *different* light, how to put your mind to the effort of reinterpreting the stories, and thence design experiments that really can change minds. Yours as well as other people's.

Obviously, while writing the book, we would like to put here some examples of truly great theories and experiments. The problem with trying to identify examples is that once they have been done, they seem so obvious. Just try to imagine what it was like to be a biologist

---

[9] Thanks to Dr Alan Thorne.

before genetics or evolution. Imagine physics before about 1900 when it was generally believed physicists had solved all their problems except a couple of trivial ones such as whether light speed was infinite or only effectively infinite, or whether matter could be divided indefinitely or was granular (quantum). Imagine physics before Newton – many will say that it wasn't physics. Or take continental drift, which seems obvious now, but in 1920 was considered absolutely absurd. Now we can measure the rate at which the New World is moving away from the Old World with lasers. Einstein's prediction that the gravity of the sun would bend light (thus making stars behind the sun visible) was considered clearly absurd as well, but was (very nearly) confirmed by observation. Now realise that in 50 years' time, modern scientists that you haven't heard of (yet) will have revolutionised your subjects to the same extent. Science destroys its history.

You will have experienced conundrums requiring lateral thinking – and you know that once you have heard the answer, you say 'Ah, of course!', and then you can't remember what it was like *not* to understand the answer. Ian Stewart's Introduction (Chapter 2) gave some examples which we hope you have been trying on your friends. You now understand about the contestant and the goats, and about 6-month versus 12-month pay increases. Here are a couple more examples to show you how you can see a very obvious part of the world in a quite different way once you have seen the point.

Question 1: The story goes that early in the 1914–18 war the Allies used all their metal to make offensive weaponry, guns and tanks, and only in 1916 did they issue troops in the trenches with those familiar tin hats. At once the number of head injuries went up by three times. Why?

Question 2: You have a bar of chocolate which has 8 x 4 squares carefully marked. You have to break the bar up into these squares, and you are allowed to break only along the marks in a straight line, one break at a time. What is the minimum number of breaks that you have to make?

If you don't see the answers immediately and obviously, enjoy sitting there not knowing. Anticipate the feeling that you will experience when you are told the answer. If you *do* see why, then are you sure and wrong?

Clues are in Chapter 9, page 108 and the answers in Chapter 12, page 155.

Science and sciencisms[1]

4

4.1 How scientists work

You have made the decision to become a scientist, even though the stereotypical scientists portrayed on television are talking heads deficient in all but intellect (even in the serious programmes like 'Horizon', never mind quiz shows and sci-fi). But you also likely suffered from your fellow students' jibes. They all realised that the spotty boffin in the comics, with pebble glasses from too much reading, who was bullied to do others' homework, and who didn't play football, became the scientist. Popularly, scientists come in two flavours: mad and dedicated.

You will have met some real scientists during your first degree, and you obviously felt that this was a worthwhile way of life, or you wouldn't be reading this book. Which are you going to be: mad or dedicated[2]. Or is the whole idea of science as viewed from popular

---

[1]According to his 4-year-old son, GFM was a 'sciencism'. Since then scientism has been used for the (contentious) belief that science has dominion over all philosophical approaches, including morals and ethics. See next footnote.

[2]You can be both, but then you are also evil, have a beautiful daughter, want to take over the world, and can only be stopped by James Bond.

culture completely mistaken? We think it is, of course, but there are many properties of science and scientists of which we are not proud.

Show most experimental scientists a beach (says JC), and they will begin counting the sand grains. We all feel that we're *doing* something if we're engaged in nice, comfortable, boring exercises. People can see you doing something if you're counting sand grains, surrounded by barrels, buckets, paper, computers, secretaries, clever little sand buggies – just imagine it! But when we're actually *thinking*, nothing comes out most of the time, and even *we* feel that the time has been wasted. Colleagues say 'He's absent-minded' when he's often actually present-minded, but with a problem elsewhere. 'Doing something' doesn't seem to include hard thinking. There is a very good case to be made, however, that all the best science has come from innovative thinking, and that the actual doing has often been done by the paid help.

There is nearly always a lot of work between the innovative thought, or even the entirely boring hypothesis, and the experimental design. Much of this work is mathematical – mathematics in its journeyman dress (not the mathematics that is queen of the sciences). It is calculating the costs of the apparatus, the ploughing of the fields for the planting of the special plots and computer time; discovering that in order to use 20 female mice every week it's a good idea to have a system which holds 1000 mice. There is usually, for those wiser workers who put quantitation to work for them *before* doing the experiments, some graphical, algebraic and topological thinking. Those who will be lucky (see Section 3.2) are those who have thought through what will be required, and have identified – and hopefully nullified – some snakes in the grass too.

This is usually not done explicitly. It is often difficult to know what calculation is useful before experiment, and many workers spend much time feeding preliminary results, or the results of other workers, into complicated statistical programs. They hope that new, or more refined, hypotheses will emerge. However, it is our experience that no hypotheses emerge that the numbers were not tortured or selected into producing. For most of us scientists, the exercise *pretends* to produce hypotheses, but is actually to make us feel more secure about the original hypothesis which drove us in the first place to find those particular papers in the literature, or to do those particular sums beforehand. Certainly, an exercise of this kind makes us feel better when we do the actual experiments. But they don't generate hypotheses. These preliminary statistical exercises (see next chapter) are in most cases absolutely essential to go *from* hypotheses *to* experimental designs (perhaps not where the question is binary: are there competent pigment cells in the white areas of black-and-white chickens? see Chapter 12, *Box 12.1*). However, people usually only use statistics to make sense out of results. It saves time, and work, to think carefully – and usually quantitatively – first.

4.2 The hierarchy of science

There is, in most people's minds, a ranking of the sciences. Physics, especially atomic physics, is among the most prestigious, while chemistry, biology, psychology and sociology are progressively 'softer'.

Mathematics seems to be the hardest, most crystalline of the scientific disciplines, and it is often argued that it forms the solid base upon which all the others rest. God, it has been said, is a geometer. Indeed, mathematics does have this crystalline structure: it is all keyed in together, so that it forms a great self-consistent whole. There are many areas of doubt, many alternatives not tested, many propositions unproven, but the structure of mathematics is like a Meccano™ or a complicated Lego™ construction, it is all of a piece.

Sciences like physics are usually thought of as having something of the same basic structure, with the middle being solid and reliable, and the edges being frontiers to which intrepid physicists are adding further pieces. This image suggests that a 'real' science, like physics or organic chemistry, is like a factory with outbuildings which will in time serve as way stations for more remote future frontiers. (We are dubious about this; it sometimes seems that when subatomic physicists invent a new particle to repair an equation or a theory, this is a 'saving' explanation – see Chapter 12. The outbuildings may be virtual reality!)

The comparable metaphor for biology, however, is usually envisaged as a ragged cloth, with a well-patterned centre at which we might find DNA, the Krebs' cycle and all the rest of the basic biochemistry, but whose ragged edges are not so much frontiers as chaotic tatters. The classification of organisms, their odd anatomies and physiologies, evolution and ecology are not, it seems, the same kind of thinking as $E = mc^2$. Philosophers of science are fond of physics, contemptuous of chemistry, bewildered by biology...[3]

Part of the reason for this is the enormous heterogeneity inherent in biological systems. It is not only that biology is a 'discovered' science, like geology or astronomy. This is unlike physics and chemistry, whose fundamentals rest on equations and other insights which have been 'invented' (and may therefore be virtual – see above). For an astronomer, a planet behaves like all planets of that kind: although all are different in internal structure and history, from the point of view

---

[3]...satisfied by psychology, muted by medicine, but happiest with history, though inarticulate about IT. Thanks to Hilary Hearnshaw.

of describing orbits they can be thought of as very similar centres of mass. If you are an atmospheric chemist, there is no such word as 'planet', but there are many different atmospheres which *are* like each other: they can be described by a few pressure words and some common compositions. If you are a parasitologist, however, there is *no* such word as parasite – you are largely concerned with differences among the variety of biological parasites, often especially among the individuals of the same species! The Inuit do not have a single word for 'snow', but (it is said) many words for different types of snow.

Here is a biological story that makes the same point, for it is a crucial concept, the ways in which we see and particularise the world about us (see also Chapter 12, where we consider the different cultural approaches to science). The story concerns an English schoolteacher who went out to New Zealand to teach in a school there. Failing to catch her class's attention with her excellent drawings of rabbits, parrots, wombats and kangaroos (she was confused about this, as many are – New Zealand, note), she began to draw a sheep. Immediately the students were riveted. They had not responded to any of the other, beautiful, lifelike creatures she'd drawn, and had given bored responses to 'What is this?' questions. But their puzzlement grew and grew as she continued her drawing of what should have been (according to *her* culture) the most familiar animal in their lives. 'What's this?' she asked. 'Surely you know what this is!' After a lot of coaxing, one of the students very timidly offered: 'Well, its rear end is a bit like a Worcester, but its front legs and quarters are more like a Merino, its face is like a Charolais, but they're much uglier, and no Charolais ever had ears like that.'

What we see, especially what we label, is *not* out there in the so-called real world; it's in our heads! Convergent scientists are the ones who organise our labelling of sheep, names for new subatomic particles and even distinguish different kinds of dreams. Their efforts lead us to believe that reality is *really* classified. Divergent scientists work to shake up the classification, because it doesn't really exist[4].

Sociology is rarely seen as a 'science' (at least, by those who know little of it), and again it's because of heterogeneity and plasticity – and because the practice changes the rules. The attitudes of people, for example, are clearly influenced by being interviewed, and may change during the interview. Such *recursion* has proved difficult for physics (in explaining quantum phenomena), and for chemistry (only recently have we realised that *most* reactions are autocatalytic). In biology the evolution or invasion of new organisms changes the environment for both others and themselves. All biology is necessarily recursive in this way. The chemists and physicists can

---

[4] We are aware of the irony of classifying scientists into two types.

get away with 'isolating' their experiments. It is far more difficult for biologists. Think about an aquarium[5]. The major difference between old biochemistry and new cell biology is that in the latter the biochemical recursions are described and are in their proper geography in the cell.

4.3 Similarities and differences

We are very good at grouping – our minds are classifiers. Everything we see is classified as similar to, or different from, everything else. Pedestrians see cars, car salesmen see Fords, Jaguars, sports cars and family saloons; car mechanics see fuel injection, six cylinders, and rusting hulks. New Zealand children don't see sheep; they see the breeds they're familiar with. As scientists, we do this as well. The lay, or schoolchild, view of a science sees great groups of subjects within it: elementary particles, organic compounds, animals.... If you get close to a subject, then you see the differences. Physicists, by and large, still contrive to see homogeneities: planets are centres of gravity for the celestial physicist. A chemist sees much more hetero-geneity, but all samples of copper sulphate are the same for her (and they might be). But biologists cannot even get identical samples of bacteria – and even if they did they wouldn't stay that way. Chemists doing biology (chemobiologists or biochemists) attempt to impose homogeneity by extracting mitochondria or other bits, and then working with them as if they were copper sulphate. They seek similarities, even between the muscle of mussels and mice.

Here is a conflict at the centre of experimental design. We need to make generalities in science: cabbages behave this way, salts of rubidium have in common that, girls with dominant fathers tend to..., etc. The more we know about science, the more we know about the differences. But much of biological advance is driven by the discovery of differences: since 1965, biologists don't see members of a species as identical, but each individual as being different in important genetic ways[6].

Sometimes scientific theories are the results of truly heroic attempts to understand deep truths. Often they are just 'lies to children' told to answer the continual 'why' questions[7]. For example, working out the structure and replication of DNA was a heroic result, but calling it a 'blueprint for life' is an attempt to paper over enormous complexity and heterogeneity with a glib phrase and a catchy story. Because our brains *need* to aggregate different phenomena under the same word, we are happy with scientific theories which do this: we see chairs,

---

[5] If you have never set up, or kept, an aquarium, the point here is that it can never be 'balanced': the fish/plants are growing, but most importantly *you* are part of the system.

[6] In the 1960s, Lewontin's group studied protein diversity in individual animals, rather than breeding them, to investigate their genetics. The amount of variation they found was immensely more than had been antici-pated. Lewontin, R.C. (1974) *The Genetic Basis of Evolutionary Change.* Columbia University Press, London.

[7] Stewart, I. and Cohen, J. (1997) *Figments of Reality.* Cambridge University Press, Cambridge.

trees, people – even 'weather' – as groups of similar things, and we are happy that all organisms use DNA to reproduce. This brain habit often gives us the comfort that we know what is going on, but frequently we have only invented stories we are fooling *ourselves* with. Being good at science is not to believe these stories (never say 'I know…'), or, rather, being able to invent several new scripts that give rise to the same drama. Bits of the plot, snatches of dialogue and incomplete character sketches are got from the data. The more ways that you can combine these into single consistent plays, each of them causally linked up into a persuasive story, the better you are at this bit of the scientific enterprise. The more ways you can think of the world, the more likely you are to find a way of joining the scraps together to give a completely novel view.

The general idea of experimenting, and therefore of experimental design, is that we find out things about the external world, checking our theories against the real world. But there is a good case to be made that we are *sorting out* our ideas about the world, rather than finding out new truths. Here is a classical example from philosophy of science courses. Theory: 'All swans are white!' Test: this swan is white, that swan is white, and all those swans over there are white; therefore, the theory is OK. Critic: 'Here's a black bird that looks just like a swan!' The proposer of the theory is now faced with a real problem: is the theory wrong, or does he deny it's a swan? In the real world, what he usually says is, 'That can't be a swan, it's black!' He has rearranged his ideas about the world, not checked his theory! It is common (actually, it sometimes looks as though it's compulsory) for people to use information to confirm prejudices rather than change them.

At this point, the convergent scientists will hold a conference to decide what the new rules are. We always have a homogeneity or heterogeneity choice: a choice of changing categorisation or changing theories (stories). All swans are white; a black swan cannot be a swan (*Figure 4.1*). Or *not* all swans are white. That sentence means throwing away a nice, simple theory, 'All swans are white', and complicating the mental world; we are all a bit reluctant to do that. Our teacher, drawing various animals and then her idea of a sheep, would have been reluctant to replace that idea with the much more complicated one her New Zealand pupils had. For a scientist, it has been said, tragedy is a beautiful theory destroyed by an ugly fact. The trick of the convergent scientist is that the theory can usually be saved by reclassifying the problem; that makes life a lot easier by avoiding the difficult job of finding a new hypothesis. Take, for example, the question of measurement.

Figure 4.1

A black bird - but is it a swan?

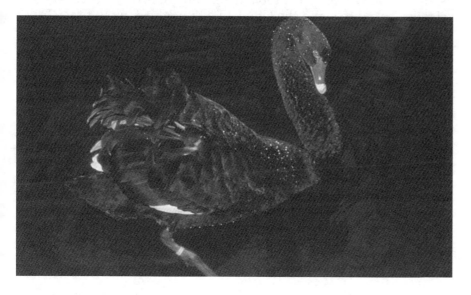

4.4 Honest measurement

Before we continue, let us make the central problem of measurement, that of accuracy and honesty, explicit. Usually, people think of measurement as having only the standard of 'accuracy'. For example, measure the height of this page. Perhaps it is 23 cm, or 9.1 inches; is it 23.1624 cm? a result that you could arrive at by making a series of independent measurements, and then taking the average on your pocket calculator, which gives you the result 'to four significant places'. Are they significant? What if you had a nine-decimal-place calculator? Or a 12-place one, so that the last figure was molecule size? Some sacks of cement suggest that you add 2.7051 litres (it used to say 4 pints) of water. If you merely want to know if the books will fit between 25-cm shelves, the best answer could be 'less than 25 cm'. There isn't a 'right' answer for the 'real' height of this page, is there?

Here is another problem: how many letters are in this line? Immediately, you have to decide if punctuation marks count as letters, if spaces are relevant. Such judgements and assessments necessarily come into every scientific observation, and it is important that you realise that you've made them – for you can't do without them! Note how different this is from swans; at least we all agree whether an individual bird is black or white[8], even if we diverge on its membership of the category 'swan'. We don't have a theory

---

[8] But Australian black swans have white patches on their backs, under the wings, which might prove confusing.

hanging on whether spaces are to count as letters, for the moment; we just want to know when to click our counter. If we count grey swans as black (or white), the problem remains simple. The convergent scientist, asked to count the letters in a line of text, would say, 'Don't confuse me with questions about punctuation marks or spaces!' But if you've been a teacher in New Zealand, or taken that lesson to heart, there are actually three, or ten, colours for swans – and you have to make decisions about punctuation marks at least, perhaps even spaces, and get three or four or five answers to the 'simple' question.

Every measurement made in science has many such decisions that it rests on. Convergent science makes these decisions. Should we measure the red shift of our new star using the old telescope, with the absorption lines marked on in pencil so we can see how much they have shifted? Or should we put in for a new model spectral analyser, which graphs everything, lists the absorption lines labelled to each chemical element, and prints out the results with the red-shift calculated. (It is said that some of the most modern ones write the scientific paper too.)

Until about 1980 most infertility clinics counted sperm 'by hand', using a haemocytometer designed for counting red blood cells against a ruled grid on a special glass microscope slide. There was a problem with this, generally unacknowledged: when sperm are sampled from a suspension, they are unlike red blood cells because they have hydro-dynamic shapes; even if they are dead they make pretty 'spray' patterns and are not evenly distributed against the grid. Modern clinics have an automated system: an undiluted drop of semen is fed into the machine, and it prints out the total concentration, percentage motile, percentage showing 'progressive motility', 'lateral head displacement' as the sperm swim, and many other data. Against this authoritative machine, which seems to have measured 'how many letters?' in many useful ways, it seems churlish to ask if the spray patterns have been taken into account. Some machines do; some don't. When measurement is automated like this, the final figure, the one you get from the machine, incorporates the prejudices of the designer, not yours! You are thereby discouraged from asking just what it is that is being measured, and that will affect what you think about the results. Once you have such a machine, you have a restricted range of questions. If you want more divergent questions you must put your other hat on, and challenge the umpires.

Here is an anecdote which underlines this. The British Antarctic Survey had primitive instruments for measuring ultraviolet light, which required calibrating each day, and only one station. Their American equivalents had several stations, with much more automated, computer-controlled equipment, which self-calibrated and dismissed outlying values. The British were the first to report the hole in the ozone layer.

Observations, examinations and experiments

5.1 Observations

Some scientists, especially astronomers, *observe*; this is the most difficult skill in science. It is only very rarely possible to know exactly what it is you're observing; and when you do, by definition almost, you can't find anything interesting, surprising or new. Sampling what you observe, or choosing a technique for examining it, has built-in problems that we can list, but which usually can't be avoided altogether even if you've identified them. We will discuss the relationship of observational uncertainty to statistical methods in Chapter 6. When critical observation is necessary, it is always as well to bear in mind one of the precepts of Rabbi Akiva: 'Take the long route which is shorter, not the short way which turns out to be much longer....' It is worth spending a long time making sure that you are familiar with what you are going to observe; you should be able to observe what other people have seen using those methods, *even if they were mistaken!* Ideally, you should be a good enough observer that you can tell why past observers *were* mistaken. Schiaparelli saw those 'canali' on Mars; any mottled pattern can produce the illusion of a network of lines, so the problem was to determine if the lines (exciting) or the mottled pattern (boring) were what you should observe. It will also be better science if you can use the apparatus at least as competently as your predecessors, and adjust and repair it if necessary. Like the apocryphal American ultra-violet light sensors, automated sampling substitutes other people's prejudices for yours, making it much more difficult to locate them. Only if you know the material thoroughly, and both the advantages and demerits of your particular piece of apparatus or software can you recognise unavoidable, built-in problems. So make sure that when you use the apparatus to observe 'standard' material you can reliably and repeatedly get the 'standard' answers. Only then will people believe your surprising, mind-changing observations.

Even then, it's a good idea to measure the same thing from a different direction. In terrestrial-based astronomy, for example, telescopes look out through a moving atmosphere, which is by no means a perfect lens. From second to second it changes its astigmatism; it has a prism effect one moment, a cylinder the next. Observing stars and planets through such an ordinary, terrestrial-based telescope shows an image that jiggles about. It has recently become possible, by establishing the star pattern with increasing accuracy, to computer-correct this jiggling, so that a nearby star can be 'fixed' with reference to the far star field. This has enabled us to resolve 'proper motion', caused by massive planets orbiting about some of those stars. We can also measure the star's movement towards and away from us, by the Doppler shift in frequency of absorption bands in its spectrum – usually the figures agree. But we could never do it persuasively through the atmosphere; only now that we have an orbit-based telescope, the Hubble, can we reduce our uncertainties to approximately the same scale as the star wobble we're trying to measure. At the same time we have improved the accuracy of our Doppler measurements, and they again agree.

The universe is unbearably complicated. Cabbages, dandelions, white and black swans, planets and stars. When we observe, or investigate, or analyse just by looking at the specimens, we have already reduced this complexity to manageable proportions – simply because eyes see only light patterns, and ignore X-rays, sound, smell, etc. Instruments like microscopes or ammeters reduce this still further. They reduce the *dimensions* of the problem. Think of this as a three-dimensional world, initially. But observation has as many dimensions as you can measure different properties. The sperm machine described above 'sees', and insists on giving you, six or seven dimensions of semen properties – each sample can differ from other samples, up or down on each of the axes, such as lateral head displacement, concentration or average progressive path length. It is embarrassing to have in the clinic if the variables it is measuring have no obvious clinical significance: that's because if you cannot assign another reason for Mr and Mrs Brown's infertility, it could be one of these variables, which you know you don't understand. A high-powered microscope reduces a tiny part of the universe to a two-dimensional picture, with colour, fluorescence and phase shift as possible other dimensions it might measure over the field. A thermometer gives us one-dimensional information, as does an ammeter. They reduce a process to a one-dimensional number – how many degrees or amps? The scale is also usually great-ly reduced – we concentrate on a small range of the possible *variation* we're investigating – our ammeter probably ignores microamps, shows milliamps on its scale, and would be destroyed by tens of amps passing through it. Our thermometer is only effective over a small range of temperature – a clinical thermometer enables us to measure only a couple of degrees above or below natural human body temperature: $38.4°C \pm {\sim}2.5°C$.

There are many kinds of observations. Where we bring the objects of our observation to an apparatus for weighing them or subjecting them to gas chromatography, we usually say we're *investigating* them. When we change them, by taking them to pieces, taking blood sam-ples or doing odd things like tests to destruction, especially if this results in lists of ingredients, factors or properties, we may say that we're *analysing* them.

One consequence of our measuring apparatus reducing dimensions is that we lose much of the complication of whatever we observe. We have to concentrate our attention on what is measurable, or perhaps assignable to one of several classes: coloured/white or even dead/alive may not be measurable. Consequently, the *variables* that relate to our question and that we are going to investigate have to be chosen from the set of things measurable (or assignable). Likewise, the *parameters* that we are going to control can only be those things that are measurable, because you cannot account for or control some-thing you can't measure – or at least, you cannot demonstrate, or claim, your mastery of its condition.

5.2 Hypothesis

As we shall see, approaching such observations with an idea in mind, a *hypothesis* about what is happening, is a necessary preliminary to any observational science. Only then do we know what we *want* to observe. (Indeed, only then do we know what to look up beforehand, what journals in the library, for example.) Do we want to weigh them, count them or shine light through them? Test their hardness, or see how far they will drive on cobbled roads until bits fall off?

The trick of a good hypothesis is that it aids in the development of understanding, by underlining what we don't understand. In particular, a good hypothesis (usually) provides a causal explanation. The best hypotheses provide causal explanations that are *unlikely*; if a hypothesis explains too much, is too easily supported by diverse evidence, it's not very satisfactory. Predilections (see Chapter 3) don't make good hypotheses. Professor Hayek (at the London School of Economics) had a powerful example of a hypothesis that is too easily supported. He used it on BBC radio to illustrate Popper's contention that analytical psychology is not science. Analytical psychology is based in Freud's theory (or belief? or hypothesis?) that juvenile emotions and frustrations are repressed into the subconscious, where they remain into adult life, affecting our attitudes and responses. Popper used *disprovability* as his criterion for 'scientific', and denied scientific status to psychoanalytical theory because it could explain anything: there *could* be no counter-examples. Hayek's story concerns a psychoanalyst who had a row with his father (or perhaps friend) about the influence of powerful fathers, and determined (or perhaps it was on a bet) that the next four patients (who happened to be women) would be explained on this basis.

1.   **Symptom** – she complained that she was sexually attracted to every male.
      Diagnosis – clearly the patient had experienced powerful paternal affection, and was looking for sex to replace it.

2.   **Symptom** – patient cannot feel sexy for anyone.
      Diagnosis – a powerful father has left so powerful an image that it excludes all other males from love and intimacy.

3.   **Symptom** – patient can feel sexy when in an intimate and private situation, but not in public.
      Diagnosis – can only feel sexy when in intimate familial surrounding, as with powerful father who modelled later relationships.

4.   **Symptom** – patient can only feel sexy in public, and cannot remain sexy in private.
      Diagnosis – men not compared to powerful father image until private (and familial) situation, and then they do not measure up.

In other words, any measurement (or symptom) – and its opposite – can be explained by the theory (about powerful fathers). And any measurement (or symptom) can, equally, be used to support an alternative explanation within the hypothesis. This kind of 'professionalism' allows us to *believe whatever we want to*. This is not science.

Much of clinical medicine is like this. People find themselves at the mercy of clinicians when they are ill. The clinician is faced with finding a reason or cause for their sickness. An important part of clinical training was called *differential diagnosis*. It was an attempt to provide clinicians with ways of helping a patient who presents with a set of symptoms, but which may suggest a number of different diseases. Differential diagnosis was about using the mind carefully and professionally. Courses in the subject demonstrated that there is not a one-to-one mapping from symptoms to diseases (or indeed from diseases to cures). Many symptoms are exhibited by several diseases, characteristic symptoms are often missing, and many diseases don't have any completely diagnostic symptoms (or, indeed, tests). Sometimes this complication is acknowledged with the label 'syndrome'[1], but even non-syndromes are more properly seen as syndromes. It is now much more common to equip the student clinician with information about a large number of technical tests which can be applied (usually to blood samples, for hormone or special peptide levels), the details of which she can't be expected to understand, to find diagnoses and, hence, cures. This progressive complexity of the clinical picture is a paradoxical reinforcement of the modern developing clinician's belief that some other speciality, or the technician in the bowels of her hospital who does the 'scientific' tests, will be able to reinforce (perhaps even provide) a definitive diagnosis. Then she can prescribe the cure associated with that diagnosis by drug-company literature. This is a way of reducing the diagnostic possibilities to one, and also the cures, without the professional involvement of the mind of the clinician.

The clinician will often align such expertise that she has with the patient's helplessness, both to be assisted by the technical answer from the tests. It does not occur to either clinician or patient, *or* to the technician doing the measurements, that the choice of tests is a beginning diagnosis – which excludes lots of possibilities. The clinician and the technician each regard the other as bedrock, and the patient is confident that, between them, they *know*. They just have to find out where this patient fits on the map. As one point on The Map.

An absolutely classical series of observations that exposes the fallacy in this method was published by Guzick *et al.*[2]. They presented infertility clinics with a series of 'infertile' couples. This was not unusual, of course. But they also randomly presented a series of 'fertile' couples (who had had children in the recent past), *labelled* infertile. The clinics did more than a dozen standard tests on each couple, and found at least one abnormal result (i.e. a reason for their infertility) in 84% of the infertile couples – fine. But they also found at least one abnormal result in 69% of the fertile couples, 'accounting for their infertility'! Given that there were only 32 matched pairs, the honest appraisal would be that about $3/4$ of all couples have clinically recognisable causes for infertility – children notwithstanding. This is an

---

[1] Simplified from 'family resemblance' to 'any 8 out of 10 symptoms' in modern medicine.

[2] Guzick, D.S., Grefenstette, I., Baffone, K., Berga, S.L., Krasnow, J.S., Stovall, D.W. and Naus, G.J. (1994) Infertility evaluation in fertile women – a model for assessing the efficacy of infertility testing. *Human Reproduction* 9, 2306–2310.

example where somebody else has presented you with a hypothesis (the patient is 'ill' – or in these cases supposedly infertile), and you *have* to find the causality. Note that the doctor's problem is solved when he has found a reason – any reason – for the infertility: 'Next patient, please.' But the patients' problem, obviously, may not be.

Many postgraduate students find themselves in this position. They are presented with a puzzle in nature, and they are required to address it as if each puzzle in nature has a one-to-one correspondence with a Ph.D. thesis. We exaggerate, but not by much – both of us have seen situations where both supervisor and student assume, or at least expect, this to be the case.

The postgraduate has to find *an* answer – any defensible answer – to the question put by his supervisor. Finding *one* apparently solves the problem. Too many Ph.D. degrees are awarded for 'getting the right answer' like this. We think that Ph.D. theses should be divergent rather than converging on the supervisor's prejudice. But then we're not convergent ourselves, as you will have noticed [3].

This is where, to some extent, the denigration of statistics comes from – 'lies, damned lies and statistics' – the absurd quip that anything can be 'proved' by abstraction and manipulation of data (even, for example, that couples with children are infertile). Good experiments (and scientists), however, should go way beyond this simplistic view of data and statistics, and one of the aims of this book is to encourage this. We are suspicious of the thesis with *one* answer to somebody else's question, just as we don't like final examination questions with an 'ideal' answer. We want to encourage you away from those simplistic directions, too, in looking for the truth. Perhaps there are religions that aspire to such certainty. The only certainty we have, as scientists, is that our hypotheses today will look as silly, in 50 years' time, as those of 50 years ago look to us now.

For those of you who have done experimental (or, indeed, any) science, this would appear to be the end of the book. Experimental design is about hypotheses, is it not? You have a hypothesis – you design the experiment to disprove it – you fail to disprove it – the hypothesis stands the test. Too many scientists, if they have the hypothesis that cabbages need phosphorus, attempt to grow cabbage plants with and without phosphorus. If the plants don't grow without phosphorus, they regard the hypothesis as proved and go on to the next one; or they or their subordinates investigate in which way phosphorus is needed, by a succession of similar *deficit experiments*.

[3] However, we have classified scientists into two groups, so we must have convergent tendencies at least.

Finding that one car in 50 drops to bits after 10 km on the cobbles, or that this particular coin comes up heads 650 times per 1000 tosses, seems to confirm, or to generate, particular hypotheses. But these lack Popperian rigour, and for advancing science are not much better than counting all the grains of sand each time you're presented with a beach. Tests of the hypotheses (rotten car! or cheat coin!) are made by *prediction* of the results of further observations[4]. (Remember, predilections don't test hypotheses; predictions should be *unlikely*.) In the simplest cases a further 500 control cars, or a further 1000 tosses of the coin could be persuasive (or, Brussels sprouts will not grow without phosphorus either). However, after you read this book we would hope that you would test the test systems themselves, by putting through *good* cars or *fair* coins that you have other information about; these have been called 'protocol experiments', and they test the tests. Such predictive, observational tests are a very usual way to proceed in regular science. Most regular scientists do deficit experiments *only* with controls (better or worse, often omitting tests of the protocol itself), as we have described[5]. We want to show you that science can – should – be much more interesting than that.

In observational sciences like palaeontology, or much of astronomy and ecology[6], it is often very difficult to persuade the universe to produce another set of the same fossils, or another Martian meteorite of the same provenance, or another beechwood with the same history, so that a prediction can be tested. Then *retrodiction*, so-called, can serve just as well: if the hypothesis holds, then these previously found fossils, or those previously analysed meteorites, will be found to have had the properties you have demonstrated. It's like finding that the make of car you tested had previously shown a very high failure rate in cobbled Provence villages or that cabbages had never successfully been grown in naturally phosphorus-deficient soils.

Observation and successful prediction is the bread and butter of science. Deficit experiments are the cheese, but can occur without the pickle of good controls. But more sophisticated experiments are the honey! We think that well-designed result-reversal experiments (Chapter 9) are the strawberries and cream.

5.3 Experiment

Experiment is usually possible in addition to observation and prediction, examining both less and more to expose the answers to simple questions. Many people think of experiments as ways to

[4] Cohen, J. and Stewart, I. (1998) That's amazing, isn't it? *New Scientist* **157:** 24–28.

[5] Three random copies of *Nature* had, at a rough count, 61 deficit experiments, 2 result reversals, 13 comparison (competition) and more than 100 observations/analyses.

[6] Hilborn, R. and Mangel, M. (1997) *The Ecological Detective*. Princeton University Press, Princeton, NJ.

restrict, or otherwise change, what we observe so that the variation we measure is more relevant to our questions. Physicists will ensure that all the physical parameters are equal, and vary only the variable of interest. Chemists will likewise ensure that temperature and pressure are equal in their various experimental situations. Biologists do the same. By inbreeding mice for many tens of generations, laboratory stocks have been achieved which have very nearly the same genes in all animals – they are like thousands of identical twins. So we can examine their susceptibility to, for example, disease without the complication of the normal genetic variation across the population.

Doing an experiment doesn't change the nature of the universe – it is still unbearably complicated, just as when making observations. By necessity, we restrict our *attentions* to a very tiny bit of the universe, and divide our little focus of interest into two groups. *Parameters* are those properties (of this micro-universe) which are not directly implicated in the hypothesis, but which we suspect can affect the results. So far as is possible, parameters are controlled within an experiment and not varied (day length, water, soil type and variety of cabbage). *Variables*, on the other hand, are those properties which are within our hypothesis and which we want to vary to test the hypothesis (phosphorus or not).

Classically, what makes an experiment an experiment is this distinc-tion, between the *control* situation and the *experimental* situation. In both situations, the parameters are maintained (controlled) so that they are as similar as we can make them. In the first (control) situation, the variables are at their normal position where the causal-ity occurs (cabbages are growing with phosphorus). In the second (experimental) situation, the variable of interest is changed to some value at which we hope to demonstrate causality, if the hypothesis is correct/useful/works (e.g. cabbages without phosphorus).

So we attempt to separate all the possible factors into those that, by and large, make no difference; those that can make a difference so they should be controlled (parameters); and those that should be varied, to see whether the result varies as they do (variables). The first two, with very difficult-to-decide boundaries between them, form the *context* of an experiment. The *content* of an experiment is the package of causality that the experimenter is trying to isolate: the action of A on X. But the *biasing* and *confounding* factors (all those B's, C's and D's) which creep in and which might affect X are also part of the content. Well-designed experiments distinguish between the effects of content and context; poorly designed experiments conflate or mix up these complications with the simple bit of causality we want to understand (and explain). However, the context defines and restricts the causality you *can* explain. An experiment involving domestic pigs and their parasites cannot provide an explanation for the effect of wing colour on predation of butterflies; nor can it provide an explanation of infection dynamics in wild boar – or can it?

Arguing outwards from an experimental result is, usually, what makes the experiment worth doing, or worth having done. Perhaps an insight from parasites in domestic pigs will explain something about wild boar. If you have thought a lot and deeply about evolution before designing the critical experiment on pigs and their parasites, it could well be that the evolutionary insight you attain will change what authors say in their introductions, then what is said in text-books, about predation of butterflies. A few very potent experiments have very wide ripples.

The *context* defines the causality structure in which your experiment is located. If you start very narrowly (investigating pig parasites for the local farmers' cooperative), it is unlikely that your results will impinge on predation defence, or even wild boar. But if you start thinking in a very wide context, there is a much greater chance that your results from pigs (which you chose as the most amenable animal model for your deep question) will give insight into apparently unrelated problems. Dare to think generally. Where, in your experiment, is the *Nature* or *Science* paper that changes lots of minds? We further address this question of context, and the meaning of scientific hypotheses and experimental results, in Chapter 7. And we return to the issue of generality when discussing sampling and populations in the next chapter.

However, no matter what effort you, as a biological scientist, put into designing experiments, remember the Harvard biological law: 'In an ideal biological experiment, with all the parameters held constant and the variables only allowed to vary within the experimental range, the organism does what it likes. Sometimes.'

What are you measuring?

6

6.1 Variability, sampling and population

Variability, sampling and *population* are the three concepts we need. These three ideas are central to the whole idea of using data to detect effects and differences. All experiments can be put in this form: you want to find differences in dependent variables between control and treatment groups for a successful experiment[1]. The central question then is, how big does the measured difference have to be before the treatment group can be considered different from the control group? For example, if the control cabbages have an average weight of 1.2 kg and treated (without-phosphorus) cabbages weigh 1.15 kg, are they different?[2] What if the treated cabbages weigh 0.5 kg on average?

The real issue is that statistics are required only when samples are involved. We are all working on populations, which, if we are careful and thoughtful, we define (or at least describe as well as we can). Choosing the population defines the *scope* of the experiment: the width of its application. It is closely related to the context issue we bought up in the previous chapter. Everything not in the experiment is in the context.

---

[1] Jumping ahead to Chapter 9, for result-reversal experiments we need two treatment groups, one of which will be demonstrably the same as the control.

[2] Note that we deliberately wrote the mean as 1.2 kg rather than 1.20 kg: we hope you noticed!

Is this make of car susceptible to cobbled roads? Clearly the population is 'this make of car'. If there is only *one* car of this type, then this is what we can experiment with. There are no statistics to be done – *the* car falls to bits or it doesn't, and we write it up accordingly. It is a binary result. If there are many of this type of car, but they are all identical in all respects, then, again, we need to look at only one of them. Once we have worked out how one particular hydrogen atom works, we can safely assume that all hydrogen atoms will behave in the same way because they are all the same. Indeed, in this case, working with hydrogen atoms is possible only if they are all the same. Of course if we didn't know about deuterium and tritium, our 'safely' might be invalidated.

But what if our car population is made up of cars that are not identical?[3] We could take just the one car and test it on cobbled roads: but that gives us only the results for the one car. How do we use that result to say all cars of this make fall to bits on cobbles? The answer, obviously, is to test more than one car. The larger the number of cars, the bigger the sample, the more our results are representative of the total population. But we are still left with the problem of transferring these results to make conclusions concerning the whole population. We want to use the results to make inferences about what would happen *if* we tested all cars. And this process of inference is statistics.

If we test the whole population (i.e. if we take a census rather than a sample), we don't need statistics. If we do an experiment and simply want to describe what happens, we don't need statistics.

- 'Fifty per cent of the infected mice in the experiment were cured with compound X' is simply a description, so no statistics are required. That's what you saw. That's what happened.

- 'Therefore, from our results we can conclude that compound X is capable of curing infected mice' is an inference, so statistics are required, as the experimental mice are taken as a sample of mice. (All mice? All mice of this genotype? All mice of this genotype from this animal house? All mice of this genotype from this animal house born in August?)

- 'Therefore, we can conclude that compound X is effective against this virus' is another inference, so statistics are required as the experimental infections (i.e. viruses) are taken as a sample of all viruses. (In mice? In all mice? In all mammals? Or going in the opposite direction, are the mice being used as samples from the population of rodents or of mammals?)

---

[3] There is a myth that cars made on Mondays and Fridays are more fault prone.

This distinction is the same as 'I tossed this coin once and it came up tails' versus 'I tossed this coin 100 times and it came up with 56 tails; therefore, it is not biased'. In the later case, the 100 tosses are a sample from all possible tosses, and we make an inference based on this sample, although we would be safer to say that it 'probably isn't biased'.

It is no coincidence that statistics was invented largely[4] to enable biologists to analyse biological data: regression to draw relationships between parental and offspring height; analysis of variance to infer something from plot-designed plant growth experiments (see Chapter 9). Because variability is the common thread of biology, and the most variability is usually found between individuals, this issue of sampling and inference is crucial to most of biology. And all the other sciences further 'downstream' where the variability rules (psychology, economics, sociology, etc.) have inherited the methods used initially within biology. Virtually all of our thinking in this area has its roots in R.A. Fisher's work with agricultural crops: it is a testimony to both the greatness of the man and the lack of imagination in the experimentalists who have followed.

In our experience, not many first-year Ph.D. students appreciate that this is the reason to do statistics. Usually they get done because everybody else does them, and/or because someone tells them to. Like wearing swimming costumes.

Deciding on, and then defining, the population and the sample should be the first steps to setting up experiments if you want them to have more than local importance; that is, if you want the scope to relate to more than the things you had in your experiment at that time. As this is not a statistics course, we will not discuss sampling methodology in any great detail. However, having defined your population, the next step should be to ensure that you do draw your sample from the whole population: that all individuals within the population have *some* chance of joining the sample. This is known as a *probability sample*. Where the chance of being selected by the experimenter is the *same* for all members of the population, we have a *random sample*.

Oh dear. Statistics (i.e. inference from your results) relies a lot on having a probability (or better, a random) sample from a population. But still (consoling hand on shoulder), the mice in your animal facility are just like all the other outbred laboratory mice in the world, so by selecting a sample of these mice you are really taking a representative sample of all outbred laboratory mice (until you discover that your animal facility mouse population was started in 1950 from a single pair of mice – survivors of a gruelling experiment involving

---

[4] Probability theory was invented for gambling, where the mechanics are laid out in front of you.

starvation and extreme cold – and fed on an unusual diet – and there was an epidemic of murine hepatitis in 1993 – and some of them are infested with mites).

This is why many laboratory animal (and especially mouse) stocks have been produced by inbreeding. C57BL6 mice really are very like other C57BL6 mice, and quite like other C57 mice. They are different from DBA or Swiss outbred mice. However, as well as the genetics, the environment, the age and the care of the animals (their husbandry) must be controlled. And this is why most modern animal facilities have a specific pathogen-free (SPF) regime. This is very expensive, but enables experiments to be at least with healthy mice of the same stock, which will be quite similar. Of course, for animals in – or from – the wild, the scientist cannot control these parameters, the genetics, their previous environments – even if the animals are cured of their natural parasites, many pathological features will remain. There will be lots of cryptic variation among wild animals, which will show up as 'noise' hiding the experimental 'signal'.

For initial laboratory experiments, it is probably enough to use cheap (undefined) mice, and only later, if the initial experiments work, use more expensive, possibly SPF, samples. For some experiments you might even need gnotobiotic mice (whose viruses and bacteria can be listed), or even axenic mice (ideally with no infection whatever). But bear in mind that mice of identical genetics usually differ *more* from each other phenotypically than outbred mice[5]. Gnotobiotic, and especially axenic, mice are *not* typical of mice, rodents or mammals – for example, their digestion is greatly impaired by poor blood flow to the gut, and they lack many endogenous vitamins. So, paradoxically, in the pursuit of a better population–sample relationship, a new population is created which, practically, bears little relationship to what we are really interested in.

We actually don't really understand why experiments use gnotobiotic or axenic mice. For most experimental designs we think this is like investigating the working of an automobile by using manual lawnmowers as models. The assertion that these animals are somehow more 'pure', more 'their DNA made flesh', is a Platonic view of the ideal organism, stripped of context. Even the use of such animals in immunology, as it were, 'the ultimate, virgin' control, avoids seeing that these virgins come with a suite of unusual characteristics (as unusual as black habits, crucifixes and Catholicism which invalidates their control status[6].)

Much modern research uses so-called SCID (severe combined immuno-deficiency) mice. Mostly they are not used as mice, but as

---

[5]A (small) 6-week-old litter of outbred mice rarely has a 10% difference between the heaviest and lightest individuals; C57BL6 inbred mice often have the heaviest weighing double the 'runt': identical mice compete more stringently for their mother's resources *in utero* and during lactation.

[6]Their immune systems have not been developed by challenge with antigens, for example; some human children, kept too clean in infancy fail to develop effective immunology (just as they would not develop language facility if isolated).

culture systems for the tissues/cancers/genes and proteins of other mammals, particularly humans. There are also many kinds of mice, which we might call 'GM mice', that have gene systems tailored for particular research purposes. Some, for example, have crippled p53 genes; these normally ensure that cells which go cancerous die by apoptosis. By inactivating that system, cancer production by these mice can be ensured and investigated without the problem of cell death. Here again, these mice must not be seen as representative of the species: people treat them as 'test tubes' for their biochemical experiments but too often fail to consider that the rest of the mouse's biochemistry is there as well. This is especially clear in those experiments that attempt to investigate the effects of individual genes by knockout procedures. The mouse, however, is more complicated than some of the thinking that goes into these experiments: most knockouts have no obvious phenotypic effects, because other genetic/developmental systems compensate. A current example is the prion protein involved in transmissible spongiform encephalopathy (as in mad cow disease), for which knockout techniques have so far found no function. But of course the knockout mice are kept in nice, warm laboratory cages, with plenty of food and water and few diseases or toxins – and their IQ is not easily measurable, or challenged.

6.2 Randomisation...

The usual, and very nearly always the best, way round the problem of inherent variability is randomisation – using chance to even things out. Let us come back to cabbages and phosphorus again. You begin, quite rightly, by randomising your cabbages between the different plots. But, little do you know that the cabbages that you have been given actually come from *two* stocks: a fast-growing variety and a slow variety (which tastes nicer – but you will never know that). To start with, let's say that only one of the 100 cabbages is a fast grower. Obviously, this one cabbage will end up in one of your plots and will either exaggerate or reduce the apparent effect of phosphorus depending on which plot it ends up in. If you are lucky, this cabbage will show up as an outlier – it will be so much bigger that you will suspect something is up. If you are not so lucky, this cabbage will contribute to the 'noise' and 'error' in your results. You will get more unexplainable variation than you anticipated: real differences might be hidden, especially if they are small.

But let us suppose that there is an equal mixture of fast and slow cabbages, i.e. 50 of each. The probability that they divide exactly evenly between the plots is small (15.8%). *Table 6.1* shows some idea of these probabilities. Perhaps the two important points to come from this table are that, first, the bigger the numbers in your experiment, the better randomisation works; and, second, that it doesn't work particularly well. Of course, the magnitude of the errors that this non-evenness of distribution will create is dependent on how big the fast/slow difference is. In this analogy, the growth-rate phenotype (fast or slow) is a *biasing* factor. Ideally, you would know about this before doing the experiment and either take it out (by making sure that all plants were of the same phenotype) or control it (by having equal numbers of each phenotype in each plot). In both cases, you are creating a parameter to reduce the amount of background noise, or inherent variability, within the experiment. Randomisation cannot reduce the noise, but it can reduce the chance that it biases the results.

Table 6.1

Probabilities associated with distribution of 'exceptions' (e.g. fast cabbages) between two plots for different numbers of exceptions and sample sizes. These probabilities can be worked out using the '@hypgeomdist' function in Excel.

Number of exceptions in sample	Sample Size	Probability all exceptions in same plot	Probability of exactly half in one plot	Probability that the distribution is not 50:50
2	100	0.495	0.505	0.495
4	100	0.117	0.383	0.617
6	100	0.027	0.322	0.678
8	100	0.006	0.285	0.715
10	100	0.001	0.259	0.741
20	100	0.000	0.197	0.803
30	100	0.000	0.172	0.828
40	100	0.000	0.162	0.838
50	100	0.000	0.158	0.842
2	50	0.490	0.510	0.490
4	50	0.110	0.391	0.609
6	50	0.022	0.333	0.667
8	50	0.004	0.298	0.702
10	50	0.001	0.275	0.725
20	50	0.000	0.227	0.773
25	50	0.000	0.214	0.786
2	20	0.474	0.526	0.474
4	20	0.087	0.418	0.582
6	20	0.011	0.372	0.628
8	20	0.001	0.350	0.650
10	20	0.000	0.344	0.656

Another example: taller people have higher IQ's[7], so if we wish to determine the effect of something on IQ we had better control for height by making sure that our control and treatment groups have the same height characteristics.

If the growth-rate phenotype is *related* to phosphorus – perhaps only the fast cabbages do better on phosphorus-rich soils and worse on deficient soils – then we have a *confounding* factor[8]. These are (by definition, unobserved) factors that are related to both the explanatory and dependent variables. The question of definition comes in because if you *know* about them, then they are not confounders. You will control for them – you will create a parameter that you measure within each group or plot. Of course, whether a factor is confounding or explanatory requires some prescience of causality or biological plausibility. Is the phosphorus/genotype interaction a nuisance creating additional noise, or is it biologically relevant to your question? Should we look at different genotypes in a separate experiment, to turn this noise into a usable variable?

Whatever the actual situation, in your experiments, it is likely that the outcome (or dependent) variable in the experiment will have some *distribution* (see *Figure 6.1* and *Boxes 6.1 and 6.2*). The whole purpose of controlling experiments by introducing parameters is to reduce the amount of variability within your measurement. This is because this variability (of unknown origin) acts as noise which can swamp the *signal* coming from the data. The signal is the message we should get from the experiment.

Figure 6.1

This figure shows a typical data distribution. It is actually a random sample from a normal distribution with mean 1.5 and standard deviation 0.5. Drawn with an Excel function '@norminv( )'

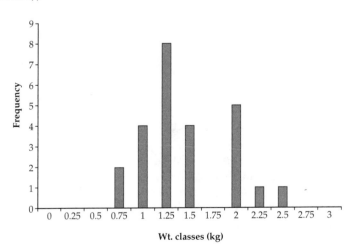

Wt. classes (kg)

---

[7] No, really, they do.

[8] A referee suggested that 'a confounder is not on the causal pathway of interest'- that we liked, and he reminded us that 'a precise and comprehensive definition can be given in terms of causal graphs' – but these are not easy to explain, we think.

Box 6.1
The 'normal' distribution

Let us suppose that the weights of 25 normal cabbages are distributed as drawn in *Figure 6.1*: three weigh between 0.25 and 0.5kg, three weigh between 0.5 and 0.75kg, and so on.

Why do the weights differ, and why are the weights distributed like that? Such a distribution can result from any or all of three situations:

1. The cabbages may be all the same size, but our measurement is equally likely to be high or low, and the probability of such error is smaller the further from the true value it is.

2. The cabbages may be of different genetic types, and by chance (or because several interbreeding generations have passed) 'dwarf' and 'giant' propensities have been randomly assigned or distributed.

3. The soil may be very varied, so that not even cabbages of similar genetics would have the same weight.

Only situation 1 gives the so-called normal distribution that is familiar to all students of biology, and which is drawn below. You would actually expect data to look like this if many measurements had been taken (not just 25 as in *Figure 6.1*). Note that the limits are actually infinite; that is there is a chance in the model of having cabbages of negative weight. The normal distribution has a mean (in this case 1.49kg) and a standard deviation (0.46kg). The standard deviation is a measure of the variability between measurements.

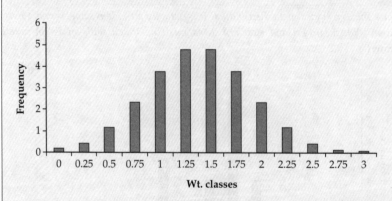

The mathematics of the normal distribution have been very well worked out, and much of statistical analysis relies on the normal distribution. However, normal distributions are better understood as standard normal distributions, since all normal distributions can be standardised (see *Box 6.3*).

There are two separable problems here, which many elementary statistics texts (and their readers) have difficulty with. Let's try to separate selection error from measurement error by inventing a very unlikely scenario. We have planted, say, 10,000 cabbages in each of two plots. Unfortunately, halfway through the experiment the plots become part of a war zone. We now have the choice of rushing out to gather the nearest 100 cabbages under cover of darkness, or we can observe the cabbages by telescope. Now, if we choose the first option, the weights could be distributed as in *Figure 6.1*. How well do these represent the 10,000?[9] The width of the telescope images is apparently related to weight, but is measurable only with real difficulty and consequent error around its real value. Let us imagine that 100 width measurements of one cabbage is, again, the figure in *Box 6.1*. Now our *measurement error* must be factored in, and it contributes to the overall variation in measurements. How do we factor it in? Think about this.

We hope you conclude that your weighing of cabbages should be supplemented by 100 weighings of one cabbage (or maybe 10); until you do this, your weight distribution is not *reliable*, because you do not know how much is measurement error and how much is natural variation. Nor is it *repeatable*, because somebody else at the balance or the telescope could get different answers. Understanding the source of the variation in your measurements (e.g. *Figure 6.1* and *Box 6.2*) is imperative if you are going to be able to interpret and use your data.

Box 6.2
The 'non-normal' distributions

Although most biologists are exposed to the normal distribution, and it underpins (for good reasons) most statistics that they use, it is actually fairly untypical. Returning to the different causes of variation in *Box 6.1*, situation 2 could give an even distribution with the same number in each box if the seeds were evenly distributed between cabbage varieties. Situation 3 would likely give an uneven distribution, but perhaps with clustering in the middle. However, for most purposes, and especially for analysis, the normal distribution is assumed.

If, as is usually the case, situations 2 and 3 pertain simultaneously (that is, a combination of distributions within two parameters of the experiment), then an overdispersed distribution is expected. That is a distribution with much more variability, genetically small cabbages in poor soil and large cabbages in good soil forming the extremes. This overdispersion is even greater if there is significant measurement

---

[9] Note that we have now turned the experimental plots of 10,000 into our 'population' so that we are actually choosing a sample from a sample.

error (situation 1). The graph below shows just such a distribution. The data (bars) are a random sample of 25 individuals from a combination of two normal distributions (two genotypes perhaps with means of 1 and 2kg and standard deviations of 0.3 and 0.7kg respectively). The line shows the sum of the distributions (that is, what we would expect with many measurements). The variation in this data is greater than in *Figure 6.1*, and has a different source.

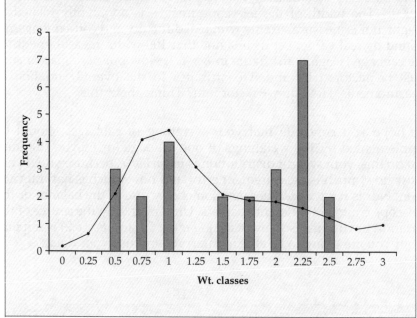

So judgements must always be made. The variability of a measurement (for example, its *standard deviation*; see *Box 6.3*) can always be reduced at a cost of time and better equipment; say, by spending another £1000. It is worth it? It is often not, for two reasons, scope and concern. If, for example, we cloned one of the cabbages, we could reduce the inherent variability (i.e. the genetic noise) to negligible levels, but we would now have results applicable only to that clone (see the discussion in Section 6.1 about laboratory animals); we would have reduced the scope. There is often such a pay-off between the scope of an experiment and the reduction in variability. However, the more important reason not to spend much more money or time to refine experiments and reduce background variability towards the beautiful zero is that your audience usually doesn't *care*. They are not concerned about your statistics, but only your conclusions. Cabbages lighter by 12% ± 6% is just as convincing to external examiners and farmers as cabbages lighter by 12.3% ± 0.6%. The latter result would typically require 100 times more cabbages and be no more persuasive.

Box 6.3
The standard normal distribution

All data can be standardised by subtracting the mean and dividing by the standard deviation. For example:

The data (1.5, 3.2, 4.3, 5.0) have a mean 3.5 and SD 1.53. Standardised, they become (-1.311, -0.197, 0.524, 0.983) with a mean of zero and a SD of one.

This is always the case. The data are now in units of SD, so that a standardised data point of 1.5 is 1.5 SD above the mean. As you might expect, multiplying the standardised data by the original SD and adding the original mean gives you back the original data. So the procedure is completely reversible.

The advantage of standardising the data is that, if the data are normally distributed, then the standard normal distribution (SND) is fully tabulated. In particular, it is known what percentage of observations are above (and below) particular values. For example, 50% of observations should be below the mean and 50% above. The graph below shows a SND as a cumulative distribution, that is, the figures on the vertical axis show the percentage expected to be below the associated value on the horizontal axis. Note that this is a truncated picture – the theory allows values from infinitely small to infinitely large.

Two values of the SND are especially important. First, 95% of all observations are expected to fall within -1.96 and +1.96. Second, 5% of observations are expected to be below -1.64, and 5% of observation above 1.64. These are required to understand hypothesis testing and generating confidence intervals.

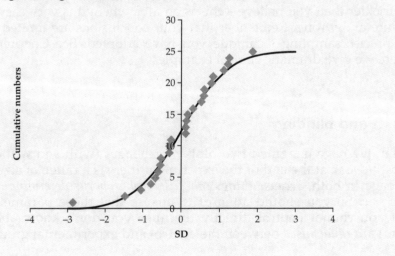

On the other hand (there always is one), an epidemiological study of the relationship between some nasty, rare cabbage disease and illness from eating these marginally lighter cabbages, where the risk of the disease is small, might require cabbage variability that *is* reliably small (or the very small-but-important signal will be lost in the noise). Depending on context, 'about half' or 'nearly all' can answer a question authoritatively. But for some questioners, 0.433 ± 0.003 is necessary. For example, 'nearly all' nuclear power stations will not explode, or 'nearly all' passengers survive air flights, do not have the authority of '0.004% ± 0.00006% of flights involve injury to passengers'[10]. *You* must decide – and justify – what accuracy is required: there isn't a textbook answer.

The important point is that you must define your population before you can sample from it properly. Each individual within the population should then have a measurable chance of entering the sample. It would be clearly fallacious to weigh a sample of 1000 people taken at random from the UK telephone directory and claim it to be a sample of all people, since most people don't have telephones … and most don't live in the UK. We must be very clear whether we are measuring people with telephones in the UK as a sample of all people, or of all people with telephones, as UK citizens – newspapers giving statistics commonly get this wrong. And there is a very important warning we should give here: don't put any reliance on historical statistics, particularly those in the media. JC was informed that 83% of the children of Israeli fighter pilots are girls and was asked, as a reproductive biologist, to explain this. It didn't need explanation; it is simply a statistical clump, like the height of women's skirts and wheat prices in 1935–55. Remember always that 'random is clumped'[11] (not evenly spread).

Interpretations and inferences that you can draw from your numbers are trickier than you believe – this is an area where it is very easy to be sure and wrong. Remember that your conclusions are limited by the precise sampling technique you have adopted. See Chapter 8, where we give dramatic clinical examples.

6.3 … and blinding

So far, we have imagined two plots of cabbages (with and without phosphorus): at the end of the experiment there is a range of size of cabbage in both, *because things other than phosphorus determine cabbage size*. So you should attempt to ensure that those parameters that you cannot control directly (because you don't know about them) are *randomised* between the control and experimental groups.

---

[10] We made these numbers up.

[11] Cohen, J. and Stewart, I. (1998) That's amazing, isn't it? *New Scientist* **157**, 24–28.

Those parameters that you *do* know about should be controlled (such as temperature and moisture).

You should also try to ensure that *you* do not know which cabbage comes from which group when you come to measure them because, especially when you get tired, you will measure them by different standards. We all do this, and the best precaution against the *bias* introduced is simply for the measurer not to know whether she's measuring a control cabbage or an experimental, phosphorus-deprived one. Getting a colleague to number the cabbages arbitrarily, and to give you the 'clues sheet' only after you've written down all their sizes, is the usual way: a random number list is useful for this.

This is standard practice for drug trials: in fact it has become almost compulsory for licensing of drugs that the trials are *double blind*, so that neither the patient nor the treating physician knows whether the patient is receiving the drug being tested[12]. The reasoning behind this is that the *placebo effect* is very strong, and people are not as independent as they like to think they are. People will, unknowingly, bias results if they have half a chance. Despite this, it is, in our experience, rare for scientists to take blinding very seriously in the laboratory. If we were joint rulers of the universe, then, papers and theses where results were obtained without blinding would *not be accepted*. It is not a criticism of your laboratory technique to adopt these procedures; it is an acknowledgement that you are human.

---

[12] So your cabbages mustn't know whether they got phosphorus or not either!

Thinking about your measurements

7.1 If you have to use statistics

If you have to use statistics, which of course you always will, then you had better be prepared to use them with the same confidence with which you can use your microscope, PCR machine and computer. The real cynic would say that you have failed to think *before* experimenting, if you have to use statistics to extract information from your results. Indeed, one of the features of good experiments is that they have been designed to give binary answers, or at least that's how it is expected that they turn out. For some questions, there really *can* only be binary answers. For example, the hypothesis that black-and-white animals do not have pigment cells in the white areas can be disproved by showing that there *are* pigment cells (as JC did in his Ph.D. thesis; see Chapter 12, *Box 12.1*). Or that finding a black swan disproves the hypothesis that all swans are white.

For most other situations, however, there *appears* to be no such obvious binary question, and the experiment usually gives a quantitative answer: for example, 50% died with treatment X and 65% with treatment Y, and you are left to decide if the treatments have different effects. This is where statistics come in, and, almost universally, are used to answer the question: did the results I get come about by chance, or do they show something systematic going on? You have to infer something about a population from a sample (*Box 7.1*).

But statistics *cannot* answer that question with a 'yes' or a 'no'; all they can do is quantify the uncertainty. Or perhaps quantify the certainty with which you can say treatment X is better than treatment Y, or A does cause X.

Box 7.1
Sampling from distributions

The graph below shows an expected distribution of data as a line (actually a negative binomial distribution for those interested – but the important point is that it isn't normal; it's overdispersed – *Box 6.2*). The mean of this population (the line) is 10. Of course, we rarely, if ever, know the population mean, and one of the aims of sampling is to estimate it. For example, to calculate the mean weight of the human population you would have to weigh everybody simultaneously.

The figure also shows a sample of 20 individuals from this population (the bars), that is, 20% (or 0.2) of the observations were zero. The mean of this sample is 11.95. In real life, this is where you start: with the sample, not the population. How does this sample mean relate to the population mean? What can we infer about the population from a particular sample, or, what can we conclude about biology from a particular experiment?

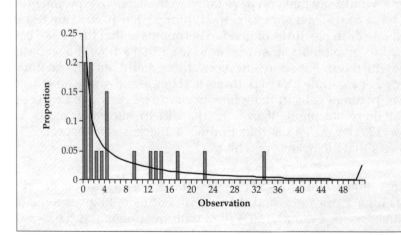

This chapter is not intended as a statistics manual[1]. There are plenty of textbooks on how to do the statistics and plenty of computer packages to do the calculations for you. What we want to do here is to get you to *think* about the statistics at a higher level – what your measurements actually *mean*, and how you interpret and explain them (especially to yourself). Most biologists use statistics[2], but largely because their supervisors and journal editors insist that they do. To continue our learning-to-swim analogy, this is like wearing swimming costumes – you are made to feel naked if you don't wear them in public baths, in the same way you are made to feel naked if you present data without statistical analyses.

The word 'statistics' has two general meanings: the discipline of the study of data, and as the plural of 'statistic' – a measurement derived from data. It is the latter meaning we take here. You have weighed your cabbages, titrated your immunoglobulin G, counted your plaques, achieved your matrix of active genes and scored your behaviours or whatever. What are you going to do with all those numbers?

The first thing, of course, is to ensure that the data are safe – send a copy to your parents at least. We used to do this. Take copies from your computer – your disk *will* crash. Keep the negatives of photographs in a fireproof safe. Some of our postgraduate students kept them in bank vaults. However, now that most images are digital, everyone knows that they are much easier to fake – make sure that the original file is kept out of your control, perhaps in an archive. A back-up is not the same as an archived version[3]. Photocopy the pages of your experimental diary and get your laboratory boss to initial them. Be inventive about methods of preventing people from thinking you have been (improperly) inventive!

The second thing is to enter the data in a spreadsheet. This has the effect of enabling you quickly to explore the data by plotting graphs every which way, which itself is an incentive to thinking about the patterns and their causes and consequences. It is easily possible to produce, for example, 3-D plots that can be dramatically informative. We are always amazed at postgraduates who, even at the stage of writing up, don't have graphs of their data plotted many different ways: 'Have you tried plotting A against X with different lines for different levels of B?' gets a blank response while the student thinks about how much time is involved doing it. Having sweated blood to get the data, make sure you

---

[1] Two particularly useful books are Kirkwood, B.R. (1988) *Essentials of Medical Statistics*, Blackwell Science; Sokal, R.R. and Rohlf, F.J. (1994) *Biometry*, W H Freeman. You can also learn much from reading (good) statistics package manuals. However, there may be better books available, or at least books that you understand better.

[2] Physicists and chemists do so, too; they are, we think, less challenged than biologists – or psychologists – to find the right statistics for their problems.

[3] Most computer systems take back-ups, and these are *not* archives.

overcome the barriers that prevent you making the best use of them[4]. At this stage you are best advised to *keep your data raw*. Don't calculate any statistics. Plot the data as they are, as you measured them. Especially, don't hide the variability by calculating standard deviation (SD) and standard error: see *Figure 7.1*. Draw the frequency distributions as in *Figure 6.1*; plot the points on graphs as they are, so that you can see *outliers, correlations* and *trends* more clearly.

Figure 7.1

The same data plotted three different ways. The first graph shows the mean plus and minus the standard deviation – all looks homogeneous. The second is a box and whisker plot showing the minima, maxima and values excluding the two highest and lowest values - downward bias is more evident. The third graph shows the raw data - clearly, there is more happening, and there could even be two groups of responses.

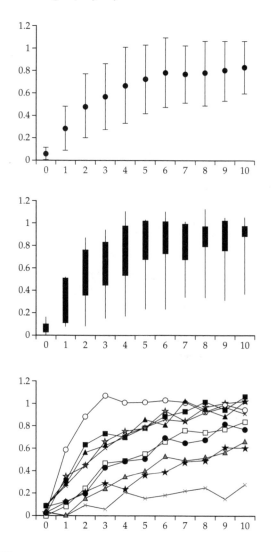

---

[4]Edward Tufte's *The Visual Display of Quantitative Information* (Cheshire, CT, 1998) has been highly recommended to us by a referee.

You may have to modify some of your results to be able to put them on the same axes as others. The telescope-determined widths of your cabbages (see above) should be cubed if you are to combine or compare them with weights of cabbages, although we hope that you would take width and weight measurements from a batch of cabbages to ensure that the conversion actually produced numbers that correlated. We are sure you see why such a *transformation* is required in this case. But others are not so obvious, and can lead to headaches in working out what you've *really* measured and what transformation is required.

7.2 Sensible statistics

Many good experimentalists find that they spend as much time analysing the data as they did collecting it. You would be wise to hope for less, but fear it may be more. However, analysis does not only mean the calculation of statistics. The first number that you will be tempted to calculate is the *arithmetic mean* (or simple average). But even this has its dangers. *Mean field theory* (in physics and mathematics) points out that in taking averages, you are assuming that the items come from the same 'field', or population. But variability (inherent in biological populations) means that things often do not come from the same field; for example, cabbages could be fast- or slow-growing; mice could be CBA or C57. Always remember, when applying mean field theory, that the 'average human being' has one testicle, one ovary, one breast and half a penis. Taking averages *across* categories produces nonsense. Worse, it hides meaningful variability; in this case, that there are two types of human being.

7.3 Different ways of showing measurement 'errors'

Virtually all numerical results published in journals now come with some estimate of variability attached. Notwithstanding the idea that 'honest' measurement might determine that some of such accuracy is spurious, this can only be a good thing. One of the mysteries of reporting measurement accuracy is what it *means*. Very few people have an intuitive feel for interpreting a standard deviation of 0.042.

The average of two sets of numbers {0,0,0,1,1,1} and {0.43, 0.51, 0.57, 0.52, 0.48, 0.49} is the same, but they show a completely different pattern. Even reporting them as having means of 0.5 and standard deviations of 0.5 and 0.042, respectively, does not do them justice (forget for a moment that meaning requires context). You cannot tell from these mean±SD statistics what the series of measurements looked like, or what they meant. Two people, looking at the same data set, might see it quite differently. But your choice of statistics, and your statistical presentation, will make this impossible.

The only well-understood idea is that of a *confidence interval*: if you repeated the same sampling or measurements again and again, 95% of the confidence intervals will contain the actual (true) population value. If we treat the little data sets above as samples, then, for the second, the 95% confidence interval is 0.47–0.53. That is, the population mean lies in this interval with a probability of 0.95. This has a much more obvious interpretation than standard deviations, standard errors or anything else. In fact, we cannot see any reason for *not* including confidence intervals. If you are ever going to actually use the information, you will want to convert it to a confidence interval. However, if all you want to do is give the reader some idea of the variability of the data, you can do that much better by show-ing all the data, or by giving the range or percentiles or some such. *Box 7.2* should give you an idea of the logic behind calculation of confidence intervals.

For the first little data sample above, the 95% confidence interval is 0.1–0.9. In this case, the value of the mean and its confidence interval depend on what sort of data they are. If the data are presence/absence data (e.g. 1 indicates the presence of a species in a location and 0 represents its absence), then the mean and confidence interval are measures of frequency and are meaningful[5]. Thus, the mean indicates that the species was present 50% of the time, and the confidence interval extends from 10% to 90%. However, if your binary (or categorical) data do not represent frequencies, means and measures of variability may be less useful. Our minds are not very good at understanding data aggregated in these ways, so be aware of this. See our Bayesian examples (Chapter 8, Section 8.2, page 81).

7.4 Transformation and scaling

In many cases, what you have measured in an experiment will not be the data that you require for presentation or analysis. Most often, you will only need fairly simple transformations to make your data presentable, comparable and interpretable. The commonest kinds of transformation are plotting data on a logarithmic scale, or dividing by 1000 to allow graphical presentation. It might be that the variable that you are interested in is percentage of killing, so that you transform your counts of cells killed to a percentage of those observed in control situations. Few people have a problem with this type of transforma-tion. It can be seen as a logical way of controlling variability.

---

[5] Even though the data cannot contain other than 0's and 1's.

Box 7.2
Properties of sample statistics

If you take a 'large' sample from a population, and calculate the mean, this sample comes from a normal distribution. This is true regardless of the population. Any population, any sample (as long as it's big enough) and the mean comes from a normal distribution. Sometimes you have to believe that God is on our side.

The graph below shows the means of 500 samples (each of size 20) plotted as a frequency distribution. The mean of the sample means (think about it) is 9.75. This looks like a normal distribution, and it is. The standard deviation of this normal distribution is given by the standard error (SE), which is the standard deviation of the population (which again, you never know) divided by the square root of the sample size (in this case $\sqrt{20}$). Not surprisingly then, the variation in the means of the samples decreases as the sample size increases. Obviously, the mean weight of a sample of 10,000 people is a better approximation to the mean weight of all people than the mean of a sample of 100.

Now we know that the distribution of the sample mean comes from a normal distribution, so we can convert this to a standard normal distribution (see *Box 6.3*). So then we can say things about the distribution of sample means, such as: 50% of all sample means will be below the population mean. We can also say that 95% of all sample means will fall within the interval (−1.96*SE, 1.96*SE) around the population mean. This can be re-expressed as saying that on 95% of samples, the confidence interval will contain the true (unknown) population mean.

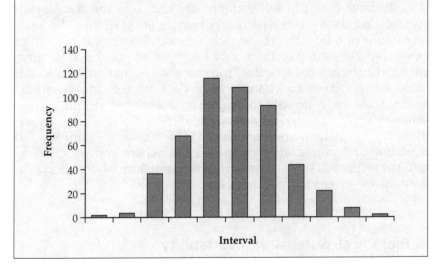

There is also a reason to standardise your data (*Box 6.3*) (standardisation being a rescaling), and that is to enable comparisons between different data sets. For example, CB mice might be inherently less variable than C57 mice in some aspect of their physiology, but standardised data will enable communicable comparisons. Commonly used IQ tests have different spreads (standard deviations); therefore, transformations are needed for psychologists to compare the measurements. For example, a person who scores 148 on the Stanford–Binet test, 2 SD from the mean (in the top 2% of the population) should score about 152 on the Cattell.

However, in terms of *analysis* of data, transformations and scaling have to be considered more carefully. In particular, the transformation has to match the requirements of the analysis. Most likely, your analysis (e.g. calculation of confidence intervals of the mean) relies on the *normal distribution* in some way, so that you should transform your data so that it has the required properties of the standard normal distribution (*Box 6.3*). Calculation of *inappropriate* standard errors and, therefore, inappropriate confidence intervals, is one of the more common errors of biologists in analysing their data. In particular, when the logical bounds of the limits are broken, this is indicative of inappropriate calculation. Examples are a percentage of killing calculated as 97% ± 4%, and the average number of cells per plate well as 3 ± 5. In both cases the calculated confidence interval contains impossible numbers (percentages over 100 and counts less than zero). The way round this is to transform the data to an appropriate scale, do the calculations and then back-transform the results to the original scale. *Box 7.3* discusses such a transformation in more detail, and gives an example of the use of logarithms on count data.

Many students (and former students) are uneasy about transforming and rescaling data – it feels slightly like cheating to take the square root of your numbers to get the right answer. Think of it as changing money: Japanese shopkeepers only take yen so you have to change your local currency to Japanese. You can always change it back again. Likewise, you have to change your data to get, for example, a sensible scale for your data to be able to use simple sampling theory – you can always change it back again. *Box 7.4* discusses a number of different transformations. This is totally unlike the calculation and presentation of mean± SD, where your original data can't be retrieved from those figures. Transformations don't lose information – they just transform it[6].

7.5 Biological systems and variability

We suggest that variability is *the* feature of biology: virtually all biological entities are *different*. The majority of biology can be couched

---

[6] And there isn't even a commission charge.

Box 7.3
Scaling and transformation of data

Suppose you have the following data, which you have gained by counting plaques or some such: 0, 0, 0, 0, 0, 0, 0, 1, 1, 2, 2, 3, 4, 4, 6, 8, 9, 11, 15, 80. You will likely have been rather alarmed at the 80 count; you will probably have checked it, and maybe even redone it.

It is tempting regard it as an 'outlier' and discard it. However, be very careful about ignoring 'aberrant' data points. They are part of the variability that you have to explain. You don't want to be in the position of having collected your swan-colour data, discard the single 'non-black' as an outlier, and then pronounce that all swans are black.

The mean of these data is 7.3, with a standard deviation of 17.64 and standard error of 3.94. If you calculate the confidence interval for the mean, it is −0.43 to 15.03. Now you are stuck, and the whole sampling theory seems to have fallen apart: you cannot have negative mean values for count data. But the theory assumes that all measurements however negative or positive are possible (Box 6.3); it hasn't failed – you have supplied it with data that are constrained to be zero or greater. So it's a simple mismatch between theory and use. If you are going to use sampling (i.e. if you are going to do almost any statistics), then you will need to make sure that the measurement could, possibly, take any value between negative and positive infinity.

You need to transform your data so that the measurements you have taken conform to this theory. In the case of count data, the appropriate transformation is to take logarithms. The main justification is that you are probably most interested in relative changes in population size, and changes from 10 to 20 and 100 to 200 are the same on a logarithmic scale (they are both multiplications by two). It also has the effect of stretching out the lower values and compressing the higher values, and as very small positive numbers become large negative logarithms, it also allows the numbers to take values across the required range.

Returning to the original data above, you will note that if you try to take a logarithm of zero, it cannot be done. So we will have to rescale the data, simply by adding 1 to each point (it could be any other number: it makes no difference since we are only going to subtract it again later). So adding 1 to each number and then taking the logarithm (to base 10) gives us a rescaled and transformed data set, with mean 0.527, standard deviation 0.524 and standard error 0.118. Now recalculating the confidence interval gives us a 95% chance that the real population mean (of the log $(x+1)$ data) lies between 0.759 and 0.296. Now we simply take the antilog and subtract one so that the estimate of the population mean becomes 2.37 (95% CI 0.98 – 4.74). Note that we are now estimating the geometric mean of the data, rather than the arithmetic mean. In the case of positively skewed data such as these (or those in *Box 6.2*), the geometric mean is a better measure of the 'centre' of the data, because it is not as sensitive to very high values.

in terms of the study of heterogeneity: how different tissues are formed from a single zygote, differential responses of plants to environmental stresses, etc. Further, all organisms even of one species are different: a biochemical map developed from successive analyses of a million bacteria might show a picture true of no individual bacterium. This is most unlike physics (where all hydrogen atoms are the same), but closer to chemistry. In many social sciences, the idea that *anything* can be considered to have an average value is sometimes ludicrous. Consequently, much (or even all) of the data collected may actually be about variability.

Box 7.4
Transformations of data

There is no absolutely correct statistic in any situation. It is largely a matter of judgement and common sense. Statistics, that is, the subject applied to real data, is not a subbranch of mathematics, so do not expect there to be the same proof that particular procedures are required in particular cases. This means that you have to understand the procedures if you are going to apply them sensibly and justifiably.

Different transformations are indicated in different cases. The table below gives a brief overview of some of the commoner transformations, and when they are likely to be useful. Note that by judicious use of additions and multiplications with logarithm and logit, it is possible to convert any scale to the full (infinite) scale.

There are other reasons for transformation, other than a restriction in scale in the original data. For example, regression requires that variables be linearly related, and one (or more) of the variables may have to be transformed to make this so. Also, many statistical tests make assumptions about variation, especially that the data themselves (not just the estimated means) are normally distributed, or that standard deviations in two groups are equal. Again transforming the data can ensure that these assumptions are not violated.

Transformation	Formula	Comments
None	$y = x$	Data already on full scale
Logarithms	$y = \log(x)$	Converts data which is zero or greater to full scale. Removes positive skew
Logit (logistic)	$y = \log\left[\dfrac{x}{1-x}\right]$	Converts proportions to full scale; regression of logit-transformed data is 'logistic regression'
Addition	$y = x + a$	Where a is any number, and may be negative so that the net effect is subtraction
Multiplication	$y = ax$	Where a is any number and so includes division

Why then, is most of the analysis done about means? Would it not often make more sense to analyse the variability? Even analysis of variance (ANOVA) is actually an analysis of differences in means[7]. We suggest that the focus on the analysis of mean values derives from the statistical, historical view of variability. It was considered to be *error*, because in chemistry, and particularly in physics, considered by statisticians to be 'real' sciences, it usually is. Regression models explicitly state this by calling all variability 'error', as in use of the term 'standard error'. The apparent, underlying concept in regression is that if we could measure things exactly, they would fall on a straight line. And it's your fault if they don't. Either you have measured it wrong or you have chosen the wrong things to measure. In chemistry or physics, this is usually so, but not in biology.

Biological variability is an intrinsic part of the system and should often be included and examined explicitly. In *Box 7.5*, an undergraduate practical is described where the outcome of the experiment is, largely, the change in patterns of variability. The students who sought means didn't achieve more than average marks from us by this technique, *even though they had been asked to find a linear function* (Q_{10}: the degree by which a biological function speeds up with temperature). The example of the roundworm *Ascaris suum* in pigs (Chapter 10) is another case where variation, not the change of a mean, was the useful answer.

If you follow the idea that variation is only error to its logical conclusion, you can end up designing experiments so that this intrinsic error is reduced to an absolute minimum – after all, nobody wants error (because it's 'your fault'). But this can lead to almost absurd levels of scope reduction or expense. Without going to absurd lengths, variation will *always* be present, even in some 'real' chemical and physical systems. Indeed, some experimental designs cannot be without variation. For example, the outcome of a birth–death process, where individual entities such as immune cells are being created and dying during the activation experiment, will always produce large variation in the numbers of cells over time. And the cells will be of different ages, too.

So the key thing is to include variability in your biological thinking. Recognise the intrinsic relationship between variability in results and the scope of the experimental results. Consider whether the outcome measure you should be looking for is a change in variability rather than in means. Even when the means change, changes in variability might be as, or more, informative: phosphorus might make cabbages grow bigger, but does it make them more or less variable?

It is becoming clear, even to a few politicians, that the major effects of global warming (almost certainly associated with increase of

---

[7] The basis of ANOVA is to see whether including different means for different groups significantly reduces the overall variation within the data.

Box 7.5
Betta egg development

Embryology undergraduate students at Birmingham University Biology School from 1969 to 1985 were given the task of collecting just-fertilised Siamese fighting fish (*Betta splendens*) eggs from embracing pairs. Each pair embraces about 30 times at 2-minute intervals, dropping 2–40 eggs each time – the physical challenge is to pick them up in a long glass pipette before the male does! Students were to measure the time interval between the 2-cell and 4-cell stages at different temperatures (each student were given two temperatures in the range 20-40°C; the fish were kept at 25°C, and bred at 27-30°C). Students were asked, misleadingly, to find the Q_{10} – the ratio by which the process went faster for a 10°C increase in temperature.

Typical results from a class of 15 students looked like the data in the graph below.

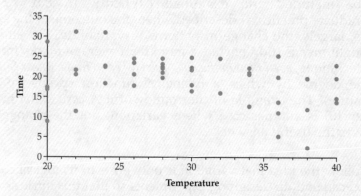

Look at the data – what do you conclude? Some students took the mean value at each temperature, and produced a wiggly line.

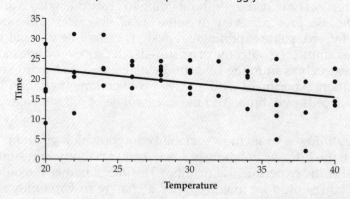

A fewer number drew a straight line (usually by eye, but we would expect regression from current students given the ease of calculation) and estimated the slope as Q_{10}. The best students noticed that the variation itself was the most interesting result. Around 25–30°C the eggs all behaved similarly – they were well buffered against change. At the extreme temperatures, near lethal at both ends, they were very variable. So the 'answer' is to plot variability _/ , and to think about canalisation (buffering of development against environmental variation).

atmospheric CO_2) are increases in the variability of weather: more floods, more droughts, more heatwaves and more severe frosts and storms, but not necessarily major changes in mean temperatures. Therefore, appropriate political responses should be made to these extreme situations occurring more frequently, rather than to difficult-to-measure mean increase of temperature.

7.6 Hypothesis testing

Why might you want to use the information imparted in confidence intervals? They are a measure of the uncertainty of measurement, but they are also ready-made hypothesis testers. If Ella reports a confidence interval and Oscar gets a result outside that interval, what does it mean? If Natasha measured the same thing 100 times, then you would expect 5 (or 5%) of her results to fall outside Ella's confidence interval, and 95 (or 95%) of the values to come within the interval. But Oscar might be measuring something different, and that's why he got a different answer. In fact, if Oscar gets even one answer outside Ella's confidence interval, he can begin to feel that he has a new story, and could even persuade an editor to publish it as a new result.

There is, in fact, no clear distinction between accurate measurement and testing of hypotheses: by delineating the amount of variability you expect from an experiment (with confidence intervals) you have already set up the limits that determine the hypothesis that another measurement comes from a different system. The question, 'Is the slope of this line greater than zero?' is the same as 'Does the confidence interval for my estimate of the slope contain the value zero?' In other words, the hypothesis that treatment A affects X is answered by measuring the effect of A on X and estimating the amount of variability expected in the measurements.

Traditionally, the experimenter is supposed to set up a *null hypothesis*, which can then be destroyed by experiment. But it is very rarely realised that the null hypothesis implicitly incorporates an assumed distribution of possible results: almost always a normal distribution. Rejecting the null hypothesis could in fact occur as a result of rejecting the *distribution*. You might have found that the distribution of results when the null hypothesis is true is not normal, rather than that the biology which generated the null hypothesis is wrong. See our discussion of whether human sperm numbers have declined for a real, important example (Chapter 8).

Investigator, beware – do not become hooked on classical hypothesis testing. Even though this is largely how experimental design and data analysis still gets taught and done. We are told to set up a:

> **Null hypothesis** = H_0: phosphorus does *not* affect cabbage growth

which usually makes the

> **Alternative hypothesis** = H_A: phosphorus *does* affect cabbage growth

Classically, you must construct such a pair of hypotheses, one of which is the logical 'not' of the other, and try to reject one. This is the see-saw model[8], and the temptation is to see it as 'left end up means right end down'. Then you must do the experiment with two groups, control and experimental. Next you test to see if the null hypothesis can be rejected. If it can, i.e. if there is a statistically significant difference between control and experimental groups, reject it: 'Phosphorus does not affect cabbage growth' is disproved.

But, the null hypothesis and its antithesis are *not* the two ends of a see-saw: you can disprove the null, but still be uninformed about the alternatives.

Especially, failure to reject the null hypothesis doesn't make it true. It is wrong to conclude that phosphorus does not affect cabbage growth, even if in your experiment it didn't. So you must distinguish between 'Phosphorus does not affect cabbage growth' and 'We can't tell from these results whether phosphorus affects cabbage growth'. If the null hypothesis is indeed true, there is indeed a high probability of actually getting the data you did. But getting this data does not provide *support* for the hypothesis: there are many, many different mechanisms that could generate your data. That you cannot reject the null hypothesis does not actually provide very strong evidence that it is true. If you fail to reject the null hypothesis, you cannot conclude that there is no difference between your control and experimental results, but only that in your experiment and its analysis a difference is not apparent. This is the main reason you need to engage your brain when you are designing experiments (see next chapter). This see-saw assumption crops up wherever the hypothesis H and the data D are seen as opposing each other, one up, t'other down! More usually, in our experience, the propositions offered to explain the data are complementary: both can be true, like the signs in Continental airports: 'Objets Trouvés[9] – Lost Property'.

Let's have another look at this opposition/complementarity:

---

[8] This is called, in some teaching systems, 'Aristotelian logic' or binary logic: proof of A is equivalent to disproof of not-A. There are many useful alternatives, with intermediate answers (like 'sometimes' or 'on 46% of occasions') or 'fuzzy' logic. Beware of concepts like 'penetrance' that import Aristotelian alternatives.

[9] Literally: 'Found Objects'.

What is the probability of obtaining a dead person (D) given that the person was hanged (H); that is, in symbol form, what is p(D | H)? Obviously, it will be very high, perhaps .97 or higher. Now, let us reverse the question. What is the probability that a person has been hanged (H) given that the person is dead (D); that is, what is p(H | D)?

This time the probability will undoubtedly be very low, perhaps .01 or lower. No one would be likely to make the mistake of sub- stituting the first estimate (.97) for the second (.01); that is, to accept .97 as the probability that a person has been hanged given that the person is dead. Even though this seems to be an unlike- ly mistake, it is exactly the kind of mistake that is made with the interpretation of statistical significance testing – by analogy, cal- culated estimates of p(H | D) are interpreted as if they were esti- mates of p(D | H), when they are clearly not the same[10].

7.7 Post-experimental statistics

This is where statistics get used most. You have done the experiment and now you want to test the results. You have been told that 95% sig- nificance is the key. You look for a significant difference, assuming that you don't have any prior knowledge (for example, a two-tailed test, where the null hypothesis is, say, that the mean is different), and it is almost significant, but not quite. Then you notice that if you had chosen a one-tailed test (where the null hypothesis is that the mean is higher, or lower), a significant result would have followed. It is not beyond the wit of most (any?) postgraduate scientists to construct an argument as to why a one-tailed test is the right one. So, not only can you bias your results by not doing the experiment blind, or selective- ly choosing data to include/exclude, but you can also do it by select- ing your statistical tests.

Generally, for this reason, results that rely too heavily on statistical tests – where *you* yourself are not convinced by the data and biologi- cal arguments – should be regarded with circumspection. Statistical significance testing really is no substitute for simpler forms of analy- sis (usually graphical inspection) and measurement of size of effects (confidence interval estimation). In the end, the significance should be scientific rather than statistical. Statistics don't persuade.

---

[10]Carver, R.P. (1978) The case against statistical testing. *Harvard Educ. Rev.* **48**, 378–399. This quotation was bor- rowed from a website compiled by David F. Parkhurst at Indiana University – www.indiana.edu/~stigtsts/. It is one of many insightful quotations deploring the emphasis of significance testing over estimation of size of effects.

7.8 Honest reporting of hypothesis testing

You have done your experiment, examined the data – paying particular attention to distributions and variability – calculated appropriate statistics (means, SD, etc.), developed confidence intervals and, finally, because it is demanded of you, tested your hypothesis. All the figures are prepared for the laboratory talk you are giving to your peers this afternoon. And you suddenly think: 'But if my head of department asks, 'But do you really believe this?' what do I say?'

You may have heard of Type I errors (my hypothesis is falsified, but it was actually true or valid) and Type II errors (my hypothesis stands, but it stands for an invalid position). As always, it is not simple[11]. This is because we cannot establish the 'truth' by any number of experiments . So our conclusions must be of the form 'I have not succeeded in disproving that A causes X, so I have built up a story about this, but of course I can't prove that A causes X.'

Type I and Type II errors are generally the same ideas as false-negative and false-positive results in a diagnostic test. Either of these can be regrettable in different circumstances. In medical situations, the desirability of making these errors changes with the consequences of the result. Thus, if a positive result leads to very unpleasant, dangerous and costly treatment, then false positives are to be avoided, even at the expense of increasing the probability of false negatives. But, if the outcome of a false negative is that a patient does not receive a safe, harmless, cheap treatment, then it is to be avoided at the expense of a false positive (when the treatment is given to somebody who wouldn't benefit from it).

Historically, science regards a false positive (A *does* cause X – when it doesn't) as a more serious error than a false negative, and, generally, biases hypothesis testing so that they are made less often than false negatives. Thus, with most experimental designs and using a 95% confidence interval, you are much more likely to leave your null hypothesis intact (when it is false), than you are to falsify it (when it is true). The point is that these errors will happen to all scientists at some time. You are going to make mistakes (errors) if you do science properly. Indeed, if you are not making errors, you are not trying enough different things. Get your head round Type I and II errors[11] and the appropriate vocabulary. Be aware of the dangers, and excitements of science – if it was easy, 'they' would have done it all by now.

---

[11] Ian Stewart tells us that there is actually a Type III error, which is getting Types I and II confused.

7.9 Pre-experimental statistics

To call in the statistician after the experiment is done may be no more than asking him to perform a post-mortem examination: he may be able to say what the experiment died of.[12]

There are two kinds of pre-experimental statistical thinking, both of which we recommend strongly. They are very different. One, the more usual (even though we fear that most scientists don't do it), is to calculate how many measurements would be necessary for the kind of uncertainty we can live with. The other is Bayesian thinking, and requires a section to itself (Section 8.2). Here's the first kind of question to ask the statistics *before* experimenting: 'How many mice will I need in each group, if the difference I'm looking for is about 10% and their SD on this variable is about 5%, to have a better-than-even chance of distinguishing the two groups?'

Better still is to evaluate the outcome before doing the work: 'If I do experiment X with 100 replicates and I get an answer of 42, how much will it change my mind?'

There are many advantages to this pre-experimental process itself. Perhaps one of the least is that, once you have obtained your data, analysis will hold few fears because you will already be familiar with it. The main advantage is that it will focus your mind on what you expect to be the outcome: if, indeed, *this* happens, then you have the comfort of knowing that you really do understand the system. But if it doesn't happen that way, what else might happen? Why? What are the possible outcomes of any similar experiment?

Because you don't know the results of the experiment before you start, you have the freedom to try anything in your pre-experimental modelling. It might be that you will give up with the experiment before you actually do it, when you see how unconvincing the results will be. This is good – try a different experimental design. What is wrong with the experiments that you are doing? You will often get the answer that you cannot get the signal (the effect you are trying to demonstrate) away from the noise. Try thinking about result reversal, about restoring function rather than depleting it (see Chapter 9).

---

[12] Ronald A. Fisher, Indian Statistical Congress, Sankhya, ca. 1938.

Here, however, we face a barrier; many (most) non-quantitative students will not be familiar enough with spreadsheets to enable this pre-experimental analysis to be done easily. However, if you do take some pre-experiment time to learn to put your imaginary experiment on a spreadsheet, it does mean that you will be happier when it comes to looking at real data. Take time to learn the basics of Excel or similar software. Go on the course that the university IT department offers every term: you will be carrying your own prospective data in your head, your own worries about how the experiment could turn out; so you'll get much more from the course than those students who only have the teacher's examples. Make up the data in a pretend laboratory book and begin the process of analysis. Virtually everything else in the world, from cars and computers to carpets and conservatories, has more time spent on design than construction, so why should your experiments be any different?

7.10 Conclusions

Statistical analysis of experimental results can tell people only exactly how uncertain you are; if the experiment has succeeded, you should be less uncertain afterwards than before. Imagine that a colleague was describing an experiment to you, an experiment just like yours – what would you want her to tell you about the numbers so that you could make up your own mind?

> Uncertainty that comes from knowledge (knowing what you don't know) is different from uncertainty coming from ignorance.[13]

That you should be designing binary-result experiments is the main point of this chapter and comes at the end. The cynic is perhaps right – if you *need* to use statistics to persuade other people, then it was a bad experimental design. A better experiment will make less use of post-experimental statistics to extract information, and more use of pre-experimental statistics to produce a result that is nearer to a conclusive or at least more persuasive demonstration. Ideally, the result will be binary, but this need not be the case even for very well-designed experiments – if it isn't, just make sure that the statistics are operating in your favour.

---

[13] Asimov, F&SF December 1994, p.116, essay 400.

Interpreting your measurements

8.1 Interpretation involves commitment

There is a popular impression of the experimental scientist as 'all head, no heart'! Intellectually committed to rationality, he (not usually a she) cannot afford prejudice as this will affect the truth of his scientific conclusions. This popular myth is about as mistaken as it's possible to be. Real scientists – top-class Nobel-Prize-winning scientists – are deeply, emotionally involved in their science; they want the result to be A causes X, and they are (briefly) heartbroken if their prejudice is shown to be mistaken – usually by them, because we scientists are constantly in the position of trying to disprove our most fondly held convictions. This mistaken picture has come from the 1920s books, *How to Be a Scientist* and *Great Experiments of the Past*, which propagated the idea that science is rational, dispassionate, honest and the most successful way of finding the truth. We were confirmed in our belief that the best scientists of the previous century, like Darwin, aspired to exactly that rational, dispassionate approach.

It is now generally acknowledged not to be like that. Indeed, with the publication of Darwin's letters we know that it was never like that, even for Darwin (although he thought that it should be). They were much more polite then, and Wallace particularly was concerned to use phrasing that apparently weighed up the concept of natural selection rather than promoting it. However, both men were committed to that view of organic evolution by natural selection as soon as they had convinced themselves.

Much of the early work on experimental design was about creating a methodology that would overcome the personality and prejudice of the person doing the science. The construction of the hypotheses and the clear delineation of the experiment are meant to remove the human aspect – the human error. You will all have been taught the exact way that experiments are to be written up: Introduction (in which the hypotheses are constructed), Method, Results, Discussion and Conclusion. It was (and still is) a great sin to put any discussion in the results or results in the discussion. JC was told by his supervisor that in a graph the points might be results, but the line was *always* discussion.

Thus, experiments have traditionally been seen as 'blank slates'. You start from where the previous experimenter ended, but your results cannot be in any way related to previous data, other than in their experimental design. If you do make such a relationship, you are putting an inappropriate (personal) view of the previous experimenter into your work. Only in that impersonal, objective way is, or was, it thought that science can proceed. Each experiment was a separate brick of knowledge – they are allowed to sit on one another, but must remain discrete and separate.

Popper created havoc, in 1932, when he produced a series of very persuasive essays that argued (and persuaded) that no scientist could start an experiment without having a prejudice about it. In order to construct the hypotheses and then design a test the scientist must have done particular reading (not just general reading) that would have given him his views. These views must have been at odds with his reading or they would not be a motive to experiment (unless he was only doing very, very safe experiments).

This methodology was supported and created by the statistical approach adopted during the last century. The traditional approach to classical hypothesis testing, discussed in the previous chapter, does not allow us to include results from outside the experiment in question. But all experiments do have information from outside – you are only testing the effect of phosphorus on cabbages because you had some previous knowledge of unfertilised cabbage patches. The classical (Fisherian) Latin squares (Chapter 9) randomisation and blinding were developed to specifically exclude bias in the results, because of the prejudices that might be imported.

However, as early as 1764 (three years after his death), a paper by the Revd Thomas Bayes was published that argued that the context of an observation was intrinsic to its interpretation. He introduced the idea that *prior* expectations can be combined with current observation, leading to modified, or *posterior*, expectations. Throughout the twentieth century this Bayesian philosophy ran parallel to the 'frequentist' approach of most applied statisticians. The two approaches are the same only if you assume that you have no prior information: the classical scientific position.

To persuade you that prior information ought to be included explicitly in your interpretation of all experimental results (including, especially, your own), here are two examples. They are dramatic because they are related to disease, but their lesson applies widely across observational and experimental science.

8.2 Bayesian thinking

We are indebted to Gigerenzer's book *Adaptive Thinking*[1] for the first two examples in this section – and indeed for the realisation that we needed to put in a section about how statistics are easily misunderstood. We found two of the examples that he gives very persuasive, and have used them here. For more of the same, go to his book. Here we are concerned to show you that, even – perhaps especially – in the very important clinical areas he presents, the counsellors get their advice very seriously wrong. It's like the contestant and the goats example in Ian Stewart's introduction: it feels just as good when you're sure and wrong.

The first example concerns young German men who, despite not having many sexual partners or injecting drugs, were diagnosed as HIV positive at a public clinic. They were deeply shocked, of course, and attended the post-test counselling. The counsellor told them that the test has a very low false-positive rate, about $1/20,000$ – so they had a 99.99995% chance, virtually a certainty, of having the disease. A significant proportion of these young men then attempted to commit suicide, and a few succeeded. This was a very serious matter, all the more so in that the counsellors were misinterpreting the measurement.

There is another measurement, an important source of data: 'How many men in this low-prevalence group actually *had* the disease?' About $1/20,000$. About the same rate as the false-positive, note. So the question was not 'How likely is the test to have failed to be correct, this

---

[1] Gigerenzer, G. (2002) *Adaptive Thinking: Rationality in the Real World.* Oxford: Oxford University Press, New York.

time, for you, Hans?' as the counsellor believed. It was much more 'Now let us think, Hans; are you the $1/20,000$ of your peers who has the disease, or are you the $1/20,000$ false-positive? I would say, as you stand before me, that you have about a 50% chance either way.' And that would have made a great difference to what the men then did. Gigerenzer sent some of his young colleagues to these counsellors, claiming to have a positive HIV test, and asking for advice. All were told that their infection was 'virtually certain'. They were then instructed to ask, 'But what about the incidence in the general population?' and were answered, 'What has that got to do with *your* case?' The counsellors couldn't, for the most part, see that they were wrong in their assessments. It can't make any difference to the odds of getting a goat or a car (Chapter 2, Section 2.3) to change your choice, can it? And the soldiers in the First War War (Chapter 3.4, Section 3.4) must have been more reckless, to increase the number of head injuries?

Here is another clinical example, about breast-cancer clinics. A woman goes to a clinic and a suspicious lump is found. A biopsy is taken, and she is told that it is 'positive'. She is very worried, of course, and she gets counselling. 'There are many kinds of lumps: cysts, and benign and malignant growths in the breast, but our assessments give false positives – tell you it's malignant when it turns out not to be – in only about one in 20 biopsies. So you have a 95% chance of having breast cancer.' Wrong again, and for the same reason (which is why we give it as another of Gigerenzer's examples). We need to know how many women who go to that clinic and are biopsied actually *have* malignancy; it turns out to be less than 1%. So the negative 99% generate a lot of false positives, about 5% (1 in 20); the less than 1% who actually *have* the disease are less than a fifth of those diagnosed with it. So the counsellor should say, 'We don't know if you're one of the false positives from the large number who *haven't* got the disease, or one of the smaller number that we get right. So you, as you stand there, have about a one-in-five chance of being a real positive!' It seems ludicrous that our chance of having a disease should be related to how many other people might have it: it's obvious, isn't it, that it's just how accurate the test is? Not on your life!

In each of these cases the prior information has very significantly modified the posterior expectation, the chance that the patient has the disease. So the test result, alone, is not very informative (although the counsellors apparently think it is totally informative). The result has to be seen in its proper context. In epidemiology, this idea is encapsulated in the 'positive predictive value' of a test. The relationship between the positive predictive value of a test and the prevalence of the condition in the population is discussed in *Box 8.1*. Of course, if the prevalence of infection is 100% in a population, the test has no additional positive predictive value at all.

Box 8.1
Predictive values

A test has a probability of giving the right answer (i.e. the sensitivity in diagnosis terms), which we can write as $P\,(\,T+\ |\ D+\,)$ – the probability of getting a test positive (T+) given that the individual is truly infected (D+). The predictive value of the test is $P\,(\,D+\ |\ T+\,)$ – the probability that the individual is diseased given that they have a positive test. The error is to treat these two things as if they are the same.

The graph shows the relationship between the true prevalence (i.e. the real chance that the individual has the disease) and the predictive value (i.e. the chance that the individual has the disease given that they have a positive test). The three lines show the relationships (from left to right) for a test with 99%, 90% and 80% sensitivity. The dotted line is for a test of 50% sensitivity – i.e. a useless test.

Note the logarithmic scale. When the true prevalence is zero, the predictive value is zero – i.e. the probability of being diseased is zero, regardless of the test result. Likewise, when the prevalence is 100%, the prediction is always that the person has the disease. Only when the disease prevalence is 50% are the test sensitivity and predictive value equal.

The same logic applies to experiments – they have a probability of getting the 'right' result, but their relationship to reality depends on the reality – the 'state of nature'. Safe, predictable experiments will almost always work, because they simply restate the state of nature. Experiments that address the unlikely are more risky, and even if very good, are unlikely to predict accurately.

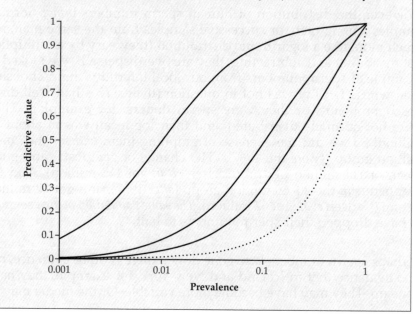

8.3 Cryptic assumptions

Sometimes the situation is much more complicated, embedded in many assumptions that haven't been examined. One such was the assertion that human sperm numbers had declined by half, between the 1930s and the 1980s. The conclusions of the original paper by Carlsen et al.[2] seemed irrefutable: the average sperm concentrations in the control men's semen, as found in published scientific papers, was about 112×10^8 in the late 1930s, and about 67×10^8 in the late 1980s. The media latched on to this, claiming that fertility was declining (for which there was no evidence at all), and many scientific groups blamed the decline on pollution of our environment with sex-hormone-mimicking chemicals. In the introductory paragraphs of that paper, the authors drew our attention to World Health Organisation (WHO) concerns in this area for many years.

Indeed, in 1967, the WHO had reduced the 'normal' sperm-number assessment from 60 million per ml to 20 million per ml. Whoops, the prior has slipped! The 'control' men before 1967 would have included only 60 million and up, whereas after 1967 'normal', suitable for control status, was 20 million and up: after 1967 the control groups had 20–60 million figures, but not previously. Could this have caused the apparent drop in numbers? Not at all likely, if a normal distribution of sperm numbers among men (and among measurements) was usual. A normal distribution had been assumed by Carlsen et al., and the omission or insertion of this group, well below the mean, would change the skewness of the graph, perhaps, but wouldn't change the mean by much. So Carlsen et al.'s paper was good, honest, important science.

However, the distribution of human sperm numbers is *not* normal; samples among men, or successive samples from the same man, are much more like a logarithmic distribution (they vary by a multiplier, not by adding and subtracting; they are overdispersed). We looked at a sample of sperm numbers from our local infertility clinic, choosing men whose fertility was not in question (their wives had well-diagnosed problems, or they were sperm donors, for example)[3]. They were not normally distributed, and their log-mean was at about 60 million! So we did the exercise of graphing them *without* the 20–60 million group (*before* the 1967 WHO change of goalposts) and got a mean of 114 million per ml (Carlsen et al. got 112 million). And we graphed *with* the 20–60 million group (*after* 1967), and got 72 million per ml (Carlsen et al. got 67 million). *The same population of men* seemed to have dropped their sperm numbers to half.

Perhaps sperm numbers *have* gone down; there is other evidence, but also evidence that in Iceland and New York, for example, they have gone up. They may have become more variable. Or the media may be

[2] Carlsen, E., Giwercman, A., Keiding, N. and Skakkebaek, N.E. (1992) Evidence for decreasing quality of semen during past 50 years. *Br. Med. J.* **305**, 609–612.

[3] Bromwich, P., Cohen, J., Stewart, I. and Walker, A. (1994) Decline in sperm counts: an artefact of changed reference range of 'normal'? *Br. Med. J.* **309**, 19–22.

reporting more differences from international means now. Those research groups who got money because of the very real alarm that Carlsen *et al.*'s paper aroused, to seek anti-androgens in the environment, have found them; we would be surprised if they hadn't. But Carlsen *et al.*'s evidence does not hold up, both because they (cryptically) assumed normal distribution, and because they didn't realise how important it was (because of the log distribution) that their 'normal' samples changed their properties over that time span.

8.4 Linking your prior to your posterior[4]

What do these examples tell us? What are the issues here for experimental design, and for persuading people of your change of mind when you have the experimental results? We expect observations to come from our priors. People will find it difficult to believe very unlikely experimental results, even though those are the ones that matter scientifically.

When JC designed the rabbit-sperm retesting experiment[5], he anticipated that if any rabbit kits[6] did result from the very few recovered oviductal sperm, he would find it difficult to believe and to convince others. The likelihood, before the experiments were done, was of the same order as mutation (or confusion) in coat colour – about $1/100,000$: the mixed-litter experiment gave one white (albino, red-eyed) kit from about 60 recovered sperm, and six brown kits from about 12,500,000 unselected washed sperm. That could have been a very unlikely colour mistake of some kind, comparable in unlikelihood to that sperm ratio. But JC also had other differences between the sires: the white buck had upright ears, as did its kit, while the six brown kits had a lop-eared father, and they were half-lops. The immune status of the kits was also consistent with their parentage (but in the event it was not discriminatory). The three variables also demonstrated that the spectacular result, the white kit from 60 sperm, had not been moved from another litter.

So it's a good idea to have several convergent – but independent – pieces of evidence if you wish to convince people of an unlikely result. Putting it classically, with the null hypothesis that all sperm are equal, our chances of getting a baby from 60 ('tired') sperm mixed with 12.5 million ('fresh') sperm would be very, very small – so small as to be unbelievable. So to make the experiment believable, distinguishable from colour error or mutation, we had to have three independent measurements confirming the paternity.

---

[4] An alternative subtitle was 'Show us your posterior', but this is an old, corny joke in the Bayesian world.
[5] JC hopes this is recalled more-or-less correctly.
[6] A baby rabbit is a kit.

Equally, if what you are demonstrating with, for example, a result-reversal (see next chapter) result is that the situation is normal, you must make that normality *surprising*. You can often do this by presenting the deficit part of the experimental results in the same table, so that the phosphorus-deprived cabbages are contrasted with the – normal – phosphorus-regained ones. JC had the high (but still less than normal) numbers of the unselected sperm as background to the remarkably low number of oviductal ones; by appropriate choice of rows and columns in tables of results, you can show that your anticipation of the unlikely result embedded it within an entirely believable context. It is not sufficient to do the experiment and get the unlikely result; a successful scientist needs his colleagues to believe that this result should change minds.

8.5 Conclusions

The important point of this chapter is that you recognise that there is no such thing as independent measurement, independent observation or independent science. Priors rule; posteriors just follow them around. The degree to which the posterior is changed by your science is the degree to which you have influenced the subject.

When it comes to understanding your results, remember that you are interpreting them in the context (i.e. the prior) of your own prejudice. When it comes to presenting your results, remember that they will be interpreted in the context (i.e. the prior) of everybody's current beliefs.

Kinds of experiments

9

9.1 Here's one we prepared earlier

A plumber diagnosing faulty central heating will try it with the thermostat at different levels. A chef will bake a cake at different temperatures and find the one that gives the best results. Although these are tantalisingly close to being experiments (and both professionals might indeed claim to be 'experimenting'), we should include them in classical science only if the 'normal' situation is included as a *control* – for example, if the plumber had an identical heating system he could compare the faulty one with, or the chef baked a 'standard' cake every time he tried a different setting. Our experience is that chefs and plumbers don't usually do this, but experimental scientists should. The educational system for chefs and plumbers is largely based on the 'here's one I prepared earlier' pattern (except that more and more we just see a computer model, and a classical picture – *doing* is unfashionable).

Science is big business. More than half the scientists that have ever lived are alive now. This is a vast market for sophisticated kits. And these kits all turn a thoughtful scientist back into a plumber (amateur) or a cook. There are kits for education in biology, especially model ecosystems that run on computers for budding architects and engineers, resource managers and conservationists. There are highly technical, very reliable kits for routine examination of materials as diverse as stainless steel, blood, sewage effluent and atmospheric dust particles, anything you need to measure in a technical society and much that you don't. There are clinical kits, diagnostic tests for a great variety of diseases and for hormone or antibody levels in blood or other body fluids. There are tests for genetic abnormalities in embryos, and for what precise single-nucleotide mutations individual people have. There are kits for doing human *in vitro* fertilisation (test-tube babies), and for many cancer diagnoses.

When people are doing these tests, they *look* just like Hollywood scientists 'doing their experiments'. But they are, mostly, *not* doing science (any more than is a carpenter measuring the length of a piece of wood, or someone assessing the oil level in a car engine with the dipstick). These are technical tricks, refined plumbing like the DIY kits that don't need any soldering and just use compression joints, or the ready-recipe cakes and meals on supermarket shelves. Just as you're not a chef when you use these ready-prepared baking kits that make life so much easier, so you're not a scientist when you use someone else's kit – Alfred's development of a sophisticated device to answer Betty's question about Charlie. Medical laboratory technicians in the UK are now, we think improperly, called 'medical laboratory scientists' because they use these sophisticated kits. However professionally they learn to use them, it isn't science.

The kits, especially the diagnostic ones, do include protocol tests, and usually have positive and negative controls too. They *include* a lot of science and experimental design, but that is so that the person operating them need *not* think scientifically. You get to see the equivalents of cabbages with phosphorus, cabbages without phosphorus, and are warned about confounding factors like cabbage-white-butterfly caterpillars. But that does not mean that the person using them to measure something is doing science. She is not, because the content and the context, the experimental design, have been defined by somebody else.

We warn you that many so-called science departments in colleges, universities and especially industry have not seen this distinction. Many people are getting Ph.D. degrees simply as a result of having done a whole series of such kit tests on material suggested by the supervisor (too often supported by a drug company or equivalent). This is like calling yourself a cordon bleu chef when you've baked, prepared or laid out already-prepared food bought from supermarket shelves. If you think this is an easy, quick, almost worry-free way to get (we nearly said 'achieve') a Ph.D., you're absolutely right. You don't need to read any more of this book, either. But because you've probably bought it, and we're a bit grateful for that, we suggest you read the one-page postamble (Chapter 13). We'll be able to recognise you for what you are not.

9.2 Kinds of experiment

As soon as the investigator questions a causal hypothesis, asks 'Does A cause X?', a comparison experiment is obvious. Such a *preliminary, trial, pilot* or *test experiment* usually compares two situations, which only differ in the presence or absence – sometimes the extent – of one factor which might affect the important properties of the test material. This factor is the variable; the maintenance of all the other conditions the same in both cases makes them parameters (see Chapter 5). This type of experiment is usually really a controlled observation, like those of careful chefs or plumbers. In fact, we should emphasise again that there is really no clear distinction between observation and experiment.

Every careful observation of a puzzling or new phenomenon should be matched to similar observations of well-understood or classical material. These are the tests of experimental protocol. Can you actually find malaria parasites in a blood smear by staining it with Giemsa, before you say that this child hasn't got malaria? Does your star spectrum show absorption lines sufficiently finely that a 0.0001% shift is distinguishable, to tell if the star has 'proper motion' away from or toward us? Can I catch enough of these fish on a standard trawling trip to show their genetic diversity? Can I do fluorimetry on this brief compound of a radioactive element? As more trouble is taken to isolate the observations and simplify the comparisons, the observation becomes more and more like an experiment. The observation that A is always observed with X becomes 'A causes X'!

'*Does A cause X?*' is the classic scientific naive question; '*Does more A cause X to increase?*' is slightly more thoughtful, but still preliminary. These are the chef and the plumber with lab coats on.

9.3 Defect experiments

The plan of a defect experiment is simple. Run the system with and without A, or in the second case with more or less A, and assess the presence, absence and amount of X. If X declines or is not seen in the absence of A, one may begin to suspect that A could just possibly be involved, in some circumstances, in something to do with the presence of X. This is cabbages with or without phosphorus. Time to begin writing the grant request, to begin discussing possibilities with colleagues, but especially to begin designing some real experiments.

This isn't sophisticated science – virtually everybody does these defect trials on a regular basis, and it is part of what we call *common sense*. If a piece of electrical equipment stops working we might first try replacing the fuse with one that we know works (because we took it out of the plug of the table lamp, which works).

A few scientists have made their careers by only doing defect experiments, but most questions about nature require rather more thought. Scientists should, we think, have more weapons available than just this very blunt instrument.

9.4 Latin squares and other dances

If you pick up a book with a title containing the phrase 'experimental design', it will almost certainly have been written by a statistician, and it will be about the *analysis of variance* (ANOVA)[1].

There are many other experimental plans that work. However, most thinking in the whole area of experimental design has been restricted to 'plot' designs, originated for comparison of different crop plants in different circumstances by Sir Ronald Fisher in the 1920s (see *Figure 9.1*). ANOVA has its name because it concentrates on the differences (in agricultural output between different plots, originally), but it is not about *explaining* the differences. It is about assigning causality: although cabbages from both our plots are indistinguishable individually, perhaps the ones without phosphorus are, on average, 1.2 g lighter. This is a clear example of how applied statistics has developed over the past century: statistics has come from biology. Or more accurately, agriculture. In the 1920s, research money wasn't spent on defence, but on agriculture.

ANOVA takes a linear view of the world: lots of Latin square designs (in many dimensions, to investigate variation along many axes) to overcome the complexity and non-linearity of the real world[2]. We now have computers, and a greater understanding of non-linear systems, which should take much of Fisher's original thinking much further, but which have also shown that some of the directions that ANOVA has taken are not particularly productive. Latin squares have been the classical experimental designs, and the analyses that they spawned (various computer ANOVA kits, ANCOVA, etc.) have been carried on *ad absurdum* in terms of their complexity to the present day.

---

[1] For example, Clarke, G.M. (1980) *Statistics and Experimental Design*. London: Edward Arnold.

[2] GFM once met a man who worked with Fisher, and claimed that Fisher could think in 5 dimensions. We don't recommend trying, although it would be rather like thinking of a page full of 3D plots.

Figure 9.1 Analysis of variance designs

A	B	A	C	A
B	C	C	B	A
B	B	C	C	B

a) A simple randomised design of five replicates of three treatments: A, B and C. Note that the treatments (for example, without phosphorus, with some phosphorus and with lots of phosphorus) have been assigned randomly to each plot.

Block			
1	B	C	A
2	B	A	C
3	C	A	B
4	B	C	A

b) A randomised block design, where the blocks represent some para meter being controlled (such as nitrogen content of soil). Each block must contain one set of the treatments (A, B and C). Again, the treatments are assigned randomly.

Block	1	2	3
1	A	B	C
2	B	C	A
3	C	A	B

c) A Latin square design which randomises each treatment (A, B and C) over two parameters for which we wish to control (such as nitrogen content and pH of soil). The design is balanced, in that each treatment occurs within each level of each block (i.e. each row and column contains each of the treatments).

They are like a machine that has continually had bits added to make it work, but whose primary design is archaic; like a tractor which has had bits bolted on to make it anything from a baling machine to a racing car. As a baling machine it works, but it is much less successful as a racing car. There are, however, many other ways of designing experiments and these have their sophistication built in. They are at least cars from the start.

9.5 Result-reversal experiments

The classic more sophisticated experiment is the 'result-reversal' format[3].

We will describe early pregnancy-test methods, because they conform to the classic *serial result-reversal* picture, then look more briefly at immunofluorescence staining techniques to show the structure of *parallel* result-reversal experiments. Then we show the relations between *qualitative* and *quantitative* designs, using the classical and very widely adopted Ames carcinogen/mutagen test as our example.

Early pregnancy tests

The principle of the early pregnancy tests was simple. It was required that the presence or absence of the characteristic pregnancy protein, hCG[4], be demonstrated in the woman's urine. This protein was extracted chemically from known-positive urine (from a pregnant woman), and injected into rabbits (with some additional immunological irritant material, called an adjuvant) so that the rabbits made antibodies to that specific human protein. If some of this rabbit's serum was run carefully down the side of a test tube containing urine from a pregnant woman, a clearly visible precipitate appeared at the interface of the liquids. 'Fine,' you say, 'the test works.' But hold: if some of this rabbit's serum was run carefully down the side of another test tube containing urine from a *non-pregnant* woman, a fairly clearly visible precipitate appeared at the interface of the liquids. Oh! The rabbit had not restricted itself only to making antibodies against hCG; it had spent much of its previous life making antibodies against all kinds of other things for its own protection, usually including the bacteria commonly found in urine samples. So its other antibodies reacted against other substances that we *weren't* curious about.

This kind of thing is usually called the *'non-specificity'* of the reagents, wrongly so, because the reaction is equally specific but not directed in the way we want. This is a very common pattern in trial experiments: A leading to X is accompanied by, and *confounded* with, a whole host of B's, C's and D's, leading to a whole plethora of W's, Y's and Z's – and possibly leading to some genuine X too!

What can be done to clean up the result? Can we make Mrs M happy by confirming her pregnancy, and make Ms N happy by reporting a real negative? A result-reversal protocol will do the trick; it will exclude the confounders.

Firstly, mix approximately the right ratios of Ms N's test urine and the serum; there *will* be a precipitate, more or less. Filter it out and discard it. Now add about the right amount of hCG solution (or some real

---

[3] JC thought this was a term in very general use, and has explained it to some 30 graduate students. However, we could not find another reference to this phrase, and must suppose JC to have invented it.

[4] Human chorionic gonadotropin.

pregnant urine from Mrs P). If there is a precipitate, the original urine had *no* hCG, because there's still enough antibody left to precipitate the hCG from Mrs P's urine – Ms N *isn't* pregnant. If there is *no* precipitate, the hCG in the original urine had used up all the antibody and there is none left for further precipitation, so the original urine *had* hCG – Ms N *is* pregnant. No precipitate, pregnant; precipitate, not pregnant: result-reversal gets rid of the 'noise' in the system and provides a clean result from a messy system. This design changes the sign of the response to hCG (plus to minus) without changing the sign of the response to 'non-specific' reaction[5]. In particular, we are no longer in the position of having to quantify the amount of precipitate, i.e. to distinguish between 'some' and 'lots', which would usually require statistics. Rather we have a binary result: some precipitate or none (*Figure 9.2*).

Interestingly, by using the result-reversal design, we end up with urine in the same state as the initial sample, whether there was hCG there or not. In this way the final result acts as its own control. The end

Figure 9.2

Pregnancy testing: the classic result reversal design.

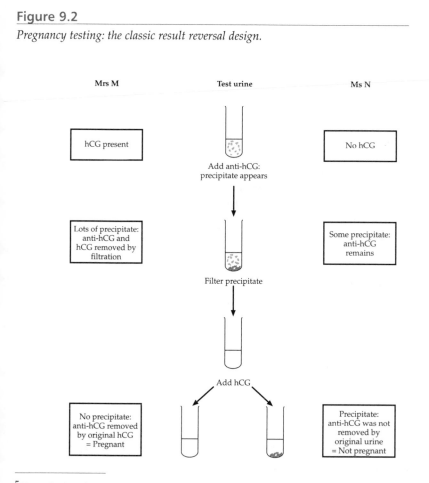

situation is a control for the start situation. In an ideal world, known-negative and known-positive urine samples would be run at the same time in every case, but this would be as a check for general experimental procedure; a protocol test, rather than for this individual result. Cost determined that it was done only for each batch of anti-hCG, or perhaps once per day, in those old pregnancy-test laboratories.

Other, much more sensitive and specific pregnancy tests are used now; most of them do still use a result-reversal format, because this gives such clean results with difficult test material.

Staining with fluorescent (or enzyme-linked) antibodies

This is a very common procedure, rarely done properly. The object is to locate a specific antigen, often a protein, in a microscope section of human or animal tissue, by causing it to react with its specific antibody. Sometimes this antibody is made fluorescent or coupled to an enzyme which can be persuaded to make a localised patch of coloured reaction product; this can then be photographed with a special microscope. More usually, however, the first antibody is made microscopically visible by coupling a second, commercially available fluorescent (or enzyme-coupled) antibody to it. Because many second antibodies can be coupled to each molecule of the first, this provides amplification and simplifies the location (and photography) of the original antigen.

Suppose we are looking for a special pathological type of actin in human muscle cells. What many laboratories do is to soak a number of microscope slides, to which sections of muscle are attached, in (diluted) rabbit immunoglobulin G (IgG) antibody prepared against the special human actin. They then rinse it off very thoroughly, so that the antibody is (hopefully) left adhering only to its specific antigen, the special actin (the manufacturers of the specific anti-special-actin antibody preparation usually tell you how much to dilute it to achieve this result). Now they soak the slides in another antibody solution, prepared in goats against rabbit IgG antibody, and heavily labelled with fluorescein. Then they wash off all the excess goat antibody and put on special mountant, and, lo, they see that the special actin is fluorescent. It usually is in such cases, and is reported as such in many published papers. But so are several other parts of the muscle cells, probably including all the normal actin. These non-specific effects allow some creativity in the observation of the results, of course. That is the way the procedure is too often done, in the form of a trial experiment. How can we clean up this system?

We need, preferably, two kinds of muscle tissue on each slide, so that we can compare normal and abnormal (they may, of course, be in the same piece of tissue). This days-of-the-week protocol was invented by

David Werrett, who went on to develop the forensic use of DNA fin-
gerprinting. It usefully separates protocol tests (Monday, Tuesday and
Friday) from control positive (Wednesday) and result reversal
(Thursday). We treated slides as follows (two of each, in case of acci-
dents – we want to be lucky, don't we?):

Monday: saline washes only, because some tissues naturally
fluoresce, and because some distilled water contains
bacterial products which fluoresce and stick to tissues,
and have been mistaken for a weak positive reaction.
This tests our microscope system's ability to detect low
levels of non-specific fluorescence, and to exclude all
but the specific fluor we're using.

Tuesday: saline wash, then goat anti-rabbit IgG, fluorescein-
labelled; *the commercial goat protein may react against –
stick to and label – those watery, bacterial products or special
actin – or react against anything else, perhaps even normal
human actin.* (JC has rejected several papers for publication because there was no way of telling whether – or
excluding that – the beautiful fluorescence was due to
the wrong reaction with the second antibody.)

Wednesday: specific rabbit anti-special actin, followed by goat anti-
rabbit IgG, fluorescein-labelled; *this is like the standard
practice – it is very difficult not to get some fluorescence with
modern high-affinity reagents.* (If nothing fluoresces on
these positive-control slides, check for heavy metals
which quench fluorescence, whether the ultraviolet
light has blown, whether the microscope is properly
aligned, etc.)

Thursday: specific rabbit anti-special actin *to which excess special
human actin has been added* to absorb the antibody, followed by goat anti-rabbit IgG, fluorescein-labelled.
This is the result-reversal test, and should look like Tuesday,
generally negative – if it looks like Wednesday,
Wednesday *is not staining special actin.*

Friday: A different rabbit-anti-human-protein (perhaps anti-*normal*-human actin) to which excess special-human-actin
has been added as in *Thursday*, followed by goat anti-
rabbit IgG, fluorescein-labelled. *Some proteins,
perhaps including our special actin, destroy or inactivate antibodies unspecifically; if our special-human actin has destroyed
the antibodies, this will show little fluorescence compared with
Wednesday, and will tell us not to believe that we have only
absorbed specifically in* Thursday.

Only if the muscle is negative in Monday, Tuesday, *and Thursday*, but positive in Wednesday (and perhaps Friday, depending what first antibody was used), has the test worked *critically*. Only then, particularly with Thursday negative when Wednesday is positive, is the presence of that very antigen confirmed. (Friday checks that the antigen is not a general protease which destroys antibodies – some proteins do.) The crucial reaction, again, is the *negative* one. Any fool can get a positive Wednesday, and many are doing so now and arguing from them. A *negative* Thursday, when the system has been absorbed with the specific antigen, is the *positive* result: result reversal again. (Incidentally, by extension of the same principle, the above experiment can be cleaned up further by absorbing *all* reagents with normal human muscle tissue, fixed to destroy proteases.)

The Ames test

In this test, we wish to know whether certain substances cause damage to the genetic material, causing either mutation or cancer (mutagenesis or carcogenesis). The problem is that many of these substances are also toxic, so that the aim of the experiment must be to separate the toxic effects from the gene-changing effects; for example, to make the gene-changing effects plus, while toxicity remains minus. Bruce Ames, in the 1950s, designed a brilliant way of doing this.

Salmonella bacteria were used, because there was a readily available mutant which had just one defect, in the synthesis path of the amino acid tryptophane. What the Ames test does is to see how often this mutation *back-mutates* so that the bacteria containing it can again live on a medium that does *not* contain tryptophane.

The procedure is to culture the bacteria on a medium *with* tryptophane, in the control situation without the suspected carcinogen. Samples are plated out to form colonies on special nutrient medium *without* tryptophane; a very few of these bacteria, which have naturally back-mutated, form colonies. About three colonies per 10,000 bacteria is common (and probably due to cosmic rays as well as the natural biogenic mutations in *Salmonella*). However, if these bacteria are grown in the presence of a mutagen such as benzpyrene (a common laboratory myth has it that this was originally extracted from the crispy bits on roast lamb) and then plated out, anything from 40 to 100 colonies per 10,000 bacteria appear. More back-mutated colonies means that the chemical tested is more mutagenic.

What this technique does is to reverse the sign for mutagenesis – less mutagenesis gives you *more* colonies growing – while retaining the sign for toxicity – more gives you *fewer* colonies growing. This is interpretable as a result-reversal experiment, but only if we realise that we are starting the experiment halfway through: the production of the tryptophane-path-deficient bacteria was the first half of the experiment (imagine that these bacteria had been incubated with the mutagen in the first place to produce the tryptophane-deficient mutant).

The cabbage experiment yet again

We have assumed, so far, that presence or absence of phosphorus was all that differed between the control and experimental plots. But think for a moment. The plot without phosphorus will probably have fewer earthworms, and it will certainly have a very different microflora and a different mineral balance. Any of these (non-specific) factors may cause a differential growth in cabbages. Can we use a result-reversal format to separate the effect of phosphorus from these other factors? We need to change the sign of the phosphorus effect, without changing the sign of the others. If we *add* phosphorus to the stunted cabbages growing in phosphorus-deficient soil, *and they grow*, then we are much surer that the difference was a direct effect of differences in phosphorus levels.

If they *don't* grow, then we have to change our minds about phosphorus acting alone: it might be a combination of phosphorus and earthworms that is required, or something more complicated. The point is that the result-reversal format for experimental design is a clever trick, in that it separates the signal from the noise. It does not replace the overriding complication of designing experiments, which is making the initial choice about what to change in the experiment, as variables, and what to hold as parameters. But it does give us a method of achieving a clean result from a designed experiment, when the initial trials (usually defect experiments) were messy to interpret or easy to misinterpret.

9.6 Demi-reversal experiments

We ought to say something about those situations where it is apparently not possible to control for everything; sometimes, indeed, for anything. We might be allowed to measure, but not to mess with the system. *Historical controls*, as one example, make statisticians shudder, but are much underrated. Experiments that use the history of the experimental material as the control – though it was not controlled at the time – may nevertheless be very persuasive. Many educational, clinical and even ecological experiments must perforce be of this kind. God may have already done the controlling for us.

Trichuris trichiura *and cognitive development*

Trichuris trichiura (the whipworm) is a ubiquitous parasite of people, especially children[6]. There has been a suggestion around for a long time of some causal association between parasite load and lack of educability, but there are all sorts of confounding factors, not least of which is that poor children are more likely to have parasites *and* to do badly at school. An elegant experiment, conducted by Kate Nokes, which overcomes much of the difficulty with experiments on people, was as follows[7]. Children were assessed for parasite load, and tested for cognitive abilities, care being taken not to associate the parasitological and educational status of each child until the end of the whole experiment. Half of the heavily infected children were then treated for worms, and all children retested.

Initially, and unsurprisingly, children with a high parasite load did worse in the tests – they were, in general, the poorest children. After treatment of some of the children, all children did better when tested, even the children who still had high parasite loads: they had all learnt how to do it. However, those children with the high parasite loads who had been treated improved substantially more than their class-mates who still had the parasites (*Figure 9.3*).

Figure 9.3

Effect of anthelminthic treatment on cognitive ability. Group C are naturally uninfected. Group T were initially infected and treated, whereas Group P were initially infected but untreated during the experiment. (Reproduced with permission of Cambridge University Press.)

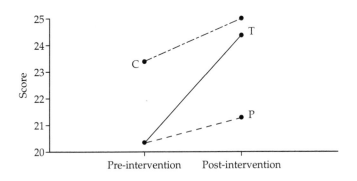

Cholera off-tap

At first glance, this does not look like an experiment at all – it looks like a repair job done by a plumber, chef or electrician. It represents,

[6] Chan, M.S. *et al.* (1994) The evaluation of global morbidity attributable to intestinal nematode infections. *Parasitology* **109**, 373–387.

[7] Nokes, C. *et al* (1992) Moderate to heavy infections of *Trichuris trichiura* affect cognitive function in Jamaican school children. *Parasitology* **104**, 539–547.

to a large extent, the founding of public health medicine, and shows how cholera is spread. John Snow, a physician, in Soho (London), after much painstaking investigation, became convinced that the water supply from one particular tap (which came from the Thames) was the source of cholera infection for local residents. In the high-handed manner of medics (nineteenth century and otherwise) he removed the tap and the incidence of cholera fell dramatically[8].There were, of course, in the strict sense, no controls in this situation. It *was* a result-reversal, in the sense that, like the worm-infected children, the population was 'returned' to a healthy state.

However, we remain unsure about these two examples. What is the normal or original state? If we accept these as the second half of a result-reversal experiment, then all defect experiments can be seen in the same way; that is, the removal/addition of a factor is the second half of the experiment, after the factor has been added/removed. The key to result-reversal is the *restoration of function*. There are possibly many ways of 'restoring' cognitive function other than removing parasites, so perhaps this is not result-reversal. There was perhaps only one way to remove cholera infection, so this might count.

There are millions of ways for a car to go wrong, to stop working. That I can stop your car from working does not mean I understand about cars. That I can *fix* your car when it has already stopped working does, indeed, suggest that I understand something about cars, because I have to have identified just that process or component which has failed.

Iron chloride in the south Atlantic

Supplementing the iron in southern Atlantic oceanic surface waters gives much more plankton production than in the same patch of ocean beforehand (i.e. historical comparison) and in nearby patches[9]. But these are not proper controls in that they are not, indeed, controlled. It is assumed that previous and nearby patches are in some sense 'normal' – God has done the controlling for us. The *other* gods, statisticians, would be at pains to point out that this is not conclusive proof, however, and that we may have witnessed a coincidence of our meddling and an algal bloom. Nonetheless, it is extremely convincing of the fact that increasing iron concentration increases algal production, despite the fact that we have not controlled iron concentration in the non-treated areas. What is really convincing is the *difference* in production between the treated and non-treated ocean surface. This is stronger evidence than the results of many clinical trials for drugs that are now accepted treatment.

---

[8] According to some accounts, the cholera level was falling anyway, and the removal of the tap was a dramatic gesture.

[9] Coates, K.H. *et al.* (1996) A massive phytoplankton bloom induced by an ecosystem-scale iron fertilization experiment in the equatorial Pacific Ocean. *Nature* **838**, 495–501.

9.7 Competition experiments

This experimental design is very common in many fields of science. We will give you three biological examples, but industrial chemistry uses this design very often, and some experiments in particle physics have this basis. The plan is rather like a result-reversal experiment, in both parallel and quantitative format. It is usually designed to test the contribution of different substances or elements to a process – the effect of deficits (as in cabbages and phosphorus) may show us some form of causality, but comparison of two or several effects can often sharpen comparisons. We will use experiments which compared the robustness of plants growing on mine tailings with their 'normal' brethren, and the competitiveness of sperm; and, for contrast, because it is more like a quantitative result-reversal experiment, the comparison of different DNAs' sequence homology with a particular DNA sequence.

Plants from mine tailings

There are many common weeds which seem to grow just as well on heavy-metal-contaminated mine refuse as on local roadsides without excess heavy metals. If these plants, or seeds from them (with some provisos – where did the pollen come from?) are planted in pots with normal soil, and roadside specimens are treated similarly, both grow apparently normally. If the same is done with contaminated soil, however, the mine-tailings' plants do all right, but the roadside plants fail badly. Why, we might ask, haven't these 'stronger', resistant plants taken over the whole population? The answer is given by the results of growing the plants crowded together in pots: the normal, roadside plants still fail in contaminated soil, but now we can see that in normal soil they overwhelm their mine-tailings congeners. The mine-tailings plants 'pay for' their special abilities by a loss in general competitiveness. Of course, one could then reduce the density, and show the recovery of out-competed plants (or rather, that's our hypothesis) in a result-reversal format.

Competition experiments are much more stringent than equivalent single tests. Heavy-metal-tolerant plants seem to grow just as well as normal sensitive wild stock when grown separately. But when they are crowded together in the same pot, without heavy metals, the wild-type, heavy-metal-sensitive plants grow better and squeeze out the metal-tolerant forms.

Acceptable and unacceptable sperm

At the end of the 1960s a new explanation for the vast numbers of spermatozoa offered by most males was suggested[10]. Because some males, as in *Drosophila*, produce very 'efficient' spermatozoa (as few as two per

[10] Cohen, J. (1967) The correlation between sperm redundancy and chiasma frequency. *Nature* **215**, 862–863.

fertilisation, but usually about 10) while man, for example, offers 3×10^8 in an ejaculate, some exponential error-generator may be implicated.

At the time, Sinclair's Cambridge pocket calculator was being produced by a process that gave only 3–7 workable circuits per 1000 attempts[11]. This was an exponential problem too, because only about half of each circuit process was successful, and therefore five such processes, $1/2 \times 1/2 \times 1/2 \times 1/2 \times 1/2$, gave only $1/32$ successful circuits – at least $31/32$ were rejected and not released to the shops. JC suggested that $1/3$ of genetic recombinations (chiasmata) might fail in some way, suggesting that the proportion of 'failed' spermatozoa should increase with the number of recombinations. *Drosophila* has no crossovers in the male, but man has about 50, suggesting that only about 2000 of 300,000,000 human sperm should be permitted to fertilise. This is a retrodiction.

In rabbits, with about 35 crossovers, about 100,000,000 spermatozoa are offered (for about 10 kits); there would be about 20,000 that should be permitted to fertilise. An experiment to test this theory needed to distinguish (1) *all spermatozoa are effectively equal, any of them could reach the finishing line (but if it's a Woody Allen style marathon, not twice in rapid succession)* from (2) *only a small proportion of these spermatozoa are permitted to fertilise, and would perhaps pass a discriminant system twice (as Sinclair's calculator circuits would pass their final quality control twice).*

It took more than two years before JC's research group could reliably recover motile sperm from rabbit oviducts to mix with fresh sperm that would produce differently coloured babies (and with other genetic differences)(see section 8.4). The mixture was then inserted into a second female. Three bucks alternated as 'selected' or 'unselected' donors. In brief, about a million unselected spermatozoa were needed per fertilisation, compared to only about 100 of those that were recovered from oviducts. The most dramatic result was a litter of one white (upright-ears) kit and six brown (lop-eared) kits from 60 oviduct-recovered, 'white' spermatozoa mixed with 12,500,000 freshly ejaculated and washed 'brown' spermatozoa. It was clear that the spermatozoa which reached the oviduct had a much better chance the second time. Traversing the female reproductive tract is neither a race nor a marathon, but a selection procedure, like the quality control in Sinclair's factory.

Tyler designed an extremely elegant experiment in the same area in the late 1970s[12]. It had been discovered that many oviductal spermatozoa did not coat over their acrosomes (the front part of their heads) with antibodies, but nearly all unselected sperm did (Werrett, who labelled IgG-labelling procedures with days of the week, had shown this). The first few sperm into the rabbit uterus (<2 hours) from the

---

[11] JC hopes this is recalled more-or-less correctly.

[12] Cohen, J. and Tyler, K.R. (1980) Sperm populations in the female genital tract of the rabbit. *J. Reprod. Fert* **60**, 213–218.

vagina were *not* coated or coatable, but later (6–8 hours) there were many more uterine sperm, almost all coated, that were attacked and eaten by white blood cells. The question was: were those oviductal sperm, which did not coat, the special fertile ones which achieved the fertilisation site again, as found in the previous experiment? And were the uncoated, first uterine sperm those which got into the oviducts quickly? Or did they coat later in the uterus, the oviductal sperm being the survivors of the uterine battle?

Tyler's beautifully economical experiment used one buck and six does, three as 'donors' and three as 'recipients' (JC's research group had used some hundreds in the reinsemination experiments). He mated the donor doe, and recovered some sperm from her uterus at about 2 hours post-copulation (600 in one case). About 60 of these sperm were spotted onto a slide for immunology; the others (at most, 540) were injected into the recipient doe (who had been injected with luteinising hormone (LH), to make her ovulate at the same time as the donor). At 8 hours, spermatozoa were recovered from the uterus of both does and spotted onto the slide (55 from the recipient, who now had 485 at most left in the reproductive tract). Ovulation occurred at about 14 hours in both does, and the oviducts were flushed to recover eggs at about 30 hours after this. The eggs were checked for fertilisation.

This experiment showed that the first spermatozoa into the donor's uterus had not coated in the recipient's uterus by 8 hours, and only about 50 were needed per fertilisation. For comparison, other experiments showed that about 20,000 ejaculate (or 8 hours uterine) unselected spermatozoa were needed per fertilisation. The first-in *were* indeed the special fertile spermatozoa which passed the female quality-control test.

These dramatic results were possible only because they built on the competition experiment protocol in JC's first experiments, when different populations were made to compete with each other.

DNA competition for DNA

Imagine three species – A, B and C – which are similar and appear to have a common, recent ancestor, and that we wish to distinguish between the relationships shown in *Figure 9.4*. The species are sufficiently alike that testing how much of A will to stick to C or B gives 100% each time: only by seeing how good each is at displacing the other can we compare them. If we mix the DNA from these species, how good is the DNA from B at preventing A from sticking to C? If we attach DNA from C to a membrane, and add equal quantities of DNA from A and B (one of them, say, A, being made radioactive or otherwise labelled), we have a competition format. The more B displaces A from sticking to C on the membrane, the less A radioactivity will be measured. Suppose that equal quantities of DNA from A and B combine with that from C (i.e. the radioactivity on the membrane is half that with A alone – or, in the real world, when A is competing with unrelated DNA, such as *E. coli* DNA). Then we must assume that A and B are identical, as measured by similarity with C. If the three species all differ equivalently, the third phylogeny is the favoured explanation. The four possibilities can be distinguished by comparison of the competitiveness of each of the three sequences with another for the third.

Figure 9.4

Four possible phylogenies and an experimental design to distinguish them.

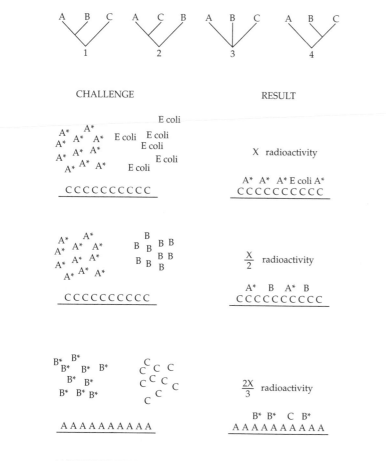

AND WHICH TREE IS FAVOURED?

9.8 The results of experiments

There are three possible kinds of result:

- *positive*, when the hypothesis is left intact (lack of phosphorus caused stunting)

- *negative*, when the hypothesis is shattered (they grew better without phosphorus)

- *null*, when the experiment hasn't worked properly.

A null is either a result which doesn't make any sense (all the cabbages explode) or an outside factor overwhelming any possible answer: all the cabbages are attacked by cabbage-white caterpillars (or all the ticks, carefully collected to assess whether they contain parasites, cook in a faulty incubator).

All scientists are taught that they should try to disprove their hypotheses, but in practice very nearly all of their experiments 'work'. They fail to disprove what they knew was right all along. So nearly all experiments give the expected, positive result. This is because much effort is put into showing that A causes X, or that it doesn't. The worker continues to refine the experiments, trying new things, as in JC's pigment-cell work (see *Box 12.1*). The experiments confirm the hypothesis, even the very clever experimental designs like Tyler's (see previous section). But, as we saw above, confirmation is not very intellectually satisfying: 'Thank goodness, another white swan!' *Nearly all 'successful' scientific experiments are tautologies* – priors are left undisturbed, adding nothing interesting to our knowledge[13]. At best, they fit present knowledge, as a brick fits in to a wall we are building or repairing. Sometimes, as in the three demi-reversal experiments above, they *are* useful for convincing other people – they make good *demonstrations*. But in each of these cases the experimenter *knew* what the result would be, and needed a way of *showing* it (we don't hear about the people who *weren't* right, who took away the wrong drinking-water tap and made no difference to cholera transmission).

If any experiment *does* show that the hypothesis is no longer tenable, there is a real logical – and political – problem. It is the converse of the idea that successful experiments produce tautologies. The problem goes like this: we have shown that A *doesn't* cause X, although we thought that it did. All the experiments we have done so far involved A (because we thought A was our causal agent), so we have no idea what *does* cause X! If only we had tried B at the same time (or noted

---

[13] But note that tautologies are still true.

how much B, C and D ... Z were in our pilot experiment). If the result is genuinely interesting and mind changing, we will discover that we made the wrong measurements.

So we ask for a new grant, or an extension, to try B. But, from the grant-giver's viewpoint, we're a poor bet. We were wrong about A, after all: B, C, D, etc. were *parameters* for our experiments, held constant for both controls and experimental set-ups. So we probably don't get approval for the grant (although the science we have done is just as groundbreaking as if we had continued to confirm that A did it!). And to add insult to injury, we can't publish our paper on 'not-A' because, firstly, everyone *knows* that A causes X, and, secondly, because it's a negative result (which many people confuse with a null). Many original scientists who do science well are, therefore, frustrated.

But do not despair; there is sociology in science too. Perhaps one of the postgraduate students in your group is dating a student of Professor B's group; what she's been doing is to show that, *in the presence of an excess of A*, B causes X (A is a parameter in her experiments). There are some figures you can use for your grant application. There is also another team, in Chicago, showing that *in the presence of adequate amounts of A and B*, C causes X. The universe is not uni-causal, there is room for more than one view.

9.9 Function deprived and restored

We suggested that the three demi-reversal situations above resemble the second half of a result-reversal experiment. In the Ames test, back-mutation restored function to the bacteria, and in the same way each of these treatments restores function to an apparently deprived situation: removing parasites 'restores' cognitive ability, removing faecally contaminated water 'restores' lack of cholera, and adding iron 'restores' ocean productivity. You don't hear about the doctor who took away a different tap or stopped people from eating the whelks, or the oceanographer who added large quantities of aluminium. In each of these well-known-because-successful cases, the experiment almost becomes a demonstration, because there is only one way to fix it. It is unique. This is like the clever mechanic who fixes the car. He does the unlikely act: he predicts, from the symptoms, that the car will operate again if he does this one little trick (perhaps cleaning the carburettor float chamber on an old car, or replacing just that one fuse on a new one). An attempt to cure a mechanical device by kicking it is attempting to exploit predilections; it's much less impressive, *even if it works*, than a successful – unlikely – prediction which works.

This is what result-reversal is about: if you can show that removing A prevents X, *and* that replacing A regenerates X, it's a much more convincing demonstration of causality than lack of A preventing X. However, even on its own, the second half (restoring function) is more convincing of causality than the first step, because there are *so many more ways* to lose function than there are to regain it.

You can only fix your car by repairing the fault, but your car can fail for a billion and one reasons.

Clues for Mind Games on page 27

Clue 1: Think about the effect of tin hats: what do they do? If you still don't see it, really feel what it is like not to understand.

Clue 2: A diagram may help, but note that this one is a trick (maybe)!

Here's the answer – what's the question?

10.1 Explanation

As a result of our observations, our investigations and our experimental results we want to be able to explain our new understanding to other people. *Explanation* is not an easy concept. At one end of the spectrum of its meanings, a teacher may accept that he has explained something to a student when she stops bothering him with questions (alas, we have all met teachers like that). At the other end, we may feel that we have to get to the point where the explainee can answer any question, from a third person, as the explainer would have done: the explainee can *substitute* for the explainer. It is not necessary that the explainee *agree* with the explainer, of course. He could say, 'If Dianne were answering you, what she would say is... . Mind you, I think she'd be wrong.' Mostly, our requirement is between these two extremes: we want the explainee to pick up some, but not all, of the explainer's understanding – and prejudices – on the subject.

Nearly all of the use of statistical methods, as discussed in Chapters 6–8, is for communication of how uncertain the experimenter is. If we tell you that the mean weight of 30 six-week-old inbred mice is 12.5g (SD 4.5g), while that of 30 six-week-old outbred mice is 15.0g (SD 2.6g), you could understand that outbred mice are not only bigger at that age, but also less variable in weight. Of course, it would be quite possible to conduct a statistical test to see if these populations are different, or rather, if we can reject the hypothesis that they are the same. Calculating the confidence intervals around the means (10.89–14.11 and 13.07–14.93) shows them to be overlapping, and suggests that we cannot reject the idea that they are the same. But even if they are different, we will say they are the same about 66% of all times that we do this sample; that is, the Type II error is 66% [1]. If they are the same, we will find them statistically different on 5% of occasions (this being the Type I error, set by the level of the significance test).

Thus, if many groups working with laboratory mice make such measurements, and all present them as papers for publication, a small percentage of them will have 'significant' results to publish, *even if there are in fact no real differences between the mice*. Hopefully, if they are not different, equal numbers of groups will claim the inbreds are heavier; but if there is some reason for interest if they are lighter, but not if they're heavier, then papers will appear claiming that inbred mice are lighter *even if they are not*. This is *reporter bias*, and although we all know it must happen, we have very little idea how common it is. JC and Ian Stewart were concerned with human sperm numbers, and whether they were declining, as had been claimed. We discovered that there was a great variety of measurements all over the world, with a very significant bias on the part of the media to quote the bad news. (The original claim for the decline had been based on a standard statistical analysis, but examination of the context suggests that this analysis is not as conclusive as was originally thought; see Section 8.3) [2].

Reporter bias is an explanation for the media misreporting human sperm numbers. Competition between genetically identical mouse babies, or outbred ones with genetically different abilities and disabilities that balance out, can explain why inbred mice are more diverse than genetically varied ones. However, we're not persuaded that understanding is needed to pass these explanations around. Perhaps the important issue is how we choose between various explanations on offer.

How many explanations are there?

Is it the case that you can explain only what you understand? [3] Prue Talbot and others had noticed that only a small proportion of the

---

[1] Calculated by assuming that the means and standard deviations given are the population values, drawing random samples from two normal distributions and using a two-tailed t-test. This can be done in any spreadsheet software. See also Berry, E.M. *et al.* (1998) The significance of non-significance. *Q. J. Med.* **91**, 647–653.

[2] Bromwich, P., Cohen, J., Stewart, I. and Walker, A. (1994) Decline in sperm counts – an artefact of changed reference range of 'normal'? *Br. Med. J.* **309**, 19–22.

[3] We are not sure about this. We leave it to the reader to discuss.

spermatozoa that reached the mass of cumulus cells around mammalian ova actually penetrated[4]. Most are stopped at the periphery of the mass, often wiggling violently but not proceeding towards the oocyte. The orthodoxy was that acrosomal hyaluronidase was required to penetrate the (hyaluronic acid-containing jelly) between the cumulus cells. Perhaps only the precisely ripe, acrosome-reacted spermatozoa had the enzyme diffusing from the 'sharp end' and could penetrate. Perhaps electron microscopy sections can show whether the ones which do penetrate have acrosome-reacted, the ones which are on the outside being too young (acrosome-complete) or senile (acrosome-lost). The 'good' ones might have holes like a pepper pot, as seen in many papers in the literature (but not, at that time, in cumulus-penetrating spermatozoa).

So, what happens with spermatozoa of different species that don't have hyaluronidase at all? Clearly, it would help if we could design an experiment that contrasts spermatozoa without hyaluronidase with penetrating and non-penetrant homologous spermatozoa, preferably with a standard mammalian egg. It turns out that neither *Xenopus* (a genus of frog) nor sea-urchin spermatozoa have hyaluronidase, so it was necessary to design a medium in which hamster sperm, hamster eggs in cumulus and *Xenopus* spermatozoa can interact – a form of competition experiment (hamster eggs are among the easiest and most robust mammalian eggs to work with). Of course, it is first necessary (protocol experiment) to check that spermatozoa without hyaluronidase *can* be negative controls, and that hamster fertilisation proceeds normally, in this medium.

Surprisingly, virtually all *Xenopus* spermatozoa penetrated the cumulus (remember that only a few homologous, hamster sperm could). If *Xenopus* spermatozoa (without hyaluronidase) do penetrate, perhaps they're special in some odd way. So Prue tried sea-urchin spermatozoa, in a sea-water-type medium (a different protocol experiment). They all penetrated too. If foreign spermatozoa can penetrate, perhaps any foreign cells can, so Prue's group tried not a sperm at all but *Chlamydomonas* algal cells (which also lack hyaluronidase). These plant single cells, swimming 'breaststroke' with two flagella, took longer, but even they penetrated 'successfully'.

This experiment was published only after a considerable number of rejections and rewrites. The results were so contrary to everybody's beliefs about the specificity of fertilisation, as shown dramatically in many other experiments. Only when Talbot saw the reprints did she see the other interpretation of these counter-intuitive results: the cumulus keeps *out* most homologous spermatozoa, the ones it lets in are 'neutral objects'. So the fertilising spermatozoa were not equipped

---

[4]Talbot, P., *et al.* (1985) Motile cells lacking hyaluronidase can penetrate the hamster oocyte cumulus complex. *Dev. Biol.* **108**, 387–398.

with a succession of special passwords, as we had all assumed; they were simply 'neutral': not 'seen'. The cumulus is *not* a guard keeping out all sperms without the correct password, but a custodian that excludes only those things that give the *wrong* password (you can decline to give any password and be accepted). JC changed his selection story in consequence of this suggestion by Talbot[5].

The non-explanation explanation: RIC!

Israeli fighter pilots were described in several newspaper reports in the 1980s as having 85% daughters, and reproductive biologists (such as JC, when he visited Israel) were challenged to provide an explanation. At the same time a local *in vitro* fertilisation clinic discovered that nearly all its boys had been produced from eggs from the left ovary, while most of the girls had come from the right ovary (it used natural, singly ovulated eggs, followed to ovulation by ultrasound). Any explanation of both oddities must include the fact that it is sperm that determine sex, not eggs.

There are in principle two kinds of explanation for this kind of observation. One is detailed, causal, and specific to each case and requires much experiment and research to verify and test. The other is to note that random does *not* mean uniform, or evenly distributed. In fact, this second explanation could be dignified with a title: RIC (random is clumped)[6]. When we look back at any history, there are all kinds of things that seem unlikely about it. It could have been that Israeli airline pilots sired all boys, or that Israeli infantrymen were all blue-eyed. Any collection of data will inevitably have many such 'clumps', all produced by ordinary random variation and unrelated to any causal mechanism. *And one in 20 of them will be statistically significant.* What, for example, would have been the probability at the start of the Jurassic period that you would be reading this book at this precise time in this precise place? What are the chances that you (as an individual) would have existed? Everything that happens (depending on the context from which it is viewed) has a vanishingly small probability of occurrence.

Science is concerned with individual causes, and with the general point that clumps happen; we need to know how to distinguish the causal issues from the statistical clump. There is, indeed, a very easy way to distinguish between a causal and a clump explanation. A causal explanation can be predicted to continue to produce the same anomalous pattern, whereas the Israeli pilots can be expected to produce half boys and half girls from now on. The explanation, in that case, is that there is nothing to explain. Significant as such patterns may look, they do indeed point to regularities in the universe, but only to the well-known property of aggregates of events that a choice few of them will do something unlikely. The more sensational media,

[5] Cohen, J. and Adeghe, A.J.-H. (1986) The other spermatozoa: fate and functions. In *New Horizons in Sperm Cell Research* (ed. H. Mohri), pp. 125–134. Japan Science Society Press, Tokyo.

[6] Cohen, J. and Stewart, I. (1998) That's amazing, isn't it? *New Scientist* **157**, 24–28.

like some tabloid newspapers, survive on the misunderstanding (by editors and readers) of this point.

Very often, the results of scientific observation or experiment do not relate to the question that the attempted explanation seems to be about. 'What has that got to do with the case in hand?' we ask. The sperm in mouse peritoneum experiments described in *Box 10.1* seemed initially to relate to the proportion of 'good' sperm in ejaculates; but the unrepeatability showed that there were other factors affecting the results. The cold-fusion experiments seemed to give information about the behaviour of hydrogen/deuterium/tritium atoms adsorbed to palladium/platinum electrodes; but nearly everyone now thinks that any surplus energy produced resulted from other, less surprising, physical or chemical events (it might be the recovery of energy used to purify the platinum/palladium electrodes, for example, as they slowly dissolve). This is often because the experiment has been badly designed, not excluding variables that affect the results but were not adequately controlled, as in the above examples. But sometimes well-designed experiments produce surprising results because such complicating factors could not be known before the experiment was done (for example, the *Ascaris suum* in pigs and the hot water freezing before cold water – see Section 10.4). This is the complement of the situation, explained in Chapter 2, that successful observations or experimental results nearly always show that other parameters should have been varied, or other variables measured in order to test the ideas we have – successful science is tautological (but note, again, that tautologies are true!).

Box 10.1
Spermatozoa in mouse peritoneum

JC's research group seemed to have found, in the late 1970s, a beautiful model for assessing how many of the 'special' fertilising spermatozoa were present in a semen sample. Spermatozoa, initially of mice, but later of known-fertile rabbits, and fertile and infertile men, were injected into the peritoneal cavities of female mice which had had a pregnancy. The sperm numbers declined to very small, but very significant, tiny numbers because most were eaten by immune defence cells. Thereafter, the numbers remained much the same for 24 hours. The surviving numbers, tens per millions injected, were beautifully correlated with the fertility of the male in most cases. But, in 1982, when a referee suggested that a parallel set of the spermatozoa should be tested in saline for percentage survival at 4 hours and 24 hours, we attempted to repeat the work. *No* spermatozoa could be recovered from the peritoneal cavity, not mouse, rabbit or human; by 4 hours *all* had disappeared. We changed mice, salines, operators and syringe makers, but have not been able to repeat these results – even though we have 350 slides with mouse-recovered spermatozoa. We have no idea what went wrong, or right.

10.2 Believability

One of the defining attributes of a scientist is scepticism. Although you might want to believe everything that you are told (and clinicians must begin by believing their patients), science will not progress if each generation believes what is in the textbooks. Each assertion has to be tested against other 'known' things (priors). If somebody claims to bend keys with 'mind power', it will cost you all your well-tested understanding of physics to believe him[7]. We should continue to bear in mind Ian Stewart's dictum that science is the best defence we have against believing what we want to.

Isaac Asimov used to give a good example of how things can be placed on a scale of believability. If a colleague comes into your office and says he has a jar of coffee in his desk, you believe him. If he says he has a bottle of uranium (and you work in a chemistry department), again, you would probably believe him. If he said uranium-235, you would be surprised, and you would probably believe him only if you had some additional evidence: he is most likely teasing you, as the screening around it would make his desk immovable. If he said astatine-218 you would not believe him, or even bother to check with anyone, as it has a half-life of 1.6 seconds and can be made only by bombarding bismuth with alpha-particles. So he's obviously fantasising – if he seems to believe it, counselling is required. You must disbelieve him, anyway, because it *can't* be true. But when you tell your supervisor about an 'unbelievable' result, you have to make it credible.

This scale of believability is altered by the perceived status of the reporter: if the person telling you is an authority figure, such as your supervisor (and therefore credible), you are more likely to believe than if it is your student. In the same way, explanations of experimental results slide along a scale of believability depending on who reports, and to whom it is being reported. It is not enough to simply have an explanation of an experimental result, you have to have a *believable* explanation; and the less senior the reporter is the more believable it must be; and the more surprising it is, the more evidence you will need to make senior listeners change their minds. Even Prue Talbot, a very senior and respected reproductive biologist, had enormous trouble getting 'fertilisation' of hamster eggs by *Xenopus* and even *Chlamydomonas* into a prestigious journal.

10.3 Hidden dimensions

When we observe, or when we investigate by experiment, we like to think that we have 'done it right'. In science, this generally means that we would like to have the confidence that other people repeating our

---

[7] You should, of course, also doubt your own scepticism: perhaps it is possible to bend keys with thought. The atom was, for a long time, the 'smallest, indivisible particle'; the Sun went round the flat Earth for many centuries.

work would get the same result (*repeatability*), and that they would get it about as often as we did if they have controlled the parameters and variables (*reliability*). Note that you would not be able to repeat the observations about Israeli fighter pilots, or right ovaries producing daughters – statistical clumping is not repeatable. We would also hope that other people would argue much the same conclusions about whether A affected X, from our (or their) set of results. However, it is rarely as simple as that; that is why the beginning of discussion sections in Ph.D. theses and scientific papers is so important, and so difficult to write. It is all about the repeatability and reliability of your results, and how much weight your results can bear.

It is too easy to cheat yourself, rather less easy to cheat others, because there is always, for a good scientist, some element of *selection of results*. Very few scientists deliberately cheat, as by inventing some points for the graph, and/or simply omitting points that don't fit. (Some idiots simply invent *all* the points; that's not science, but attempted deception – we're not concerned to tell you how to do that, in this book anyway.) It is very important to realise that selection is entirely proper; indeed, usually necessary. Some examples, close to our own experiences: you can't accept all of one week's results, because you discovered that someone else's cuvettes had been measured in the spectrophotometer, and yours had been thrown away, contaminated or relabelled; you discovered a disease in your mice which would have affected some of your stress-by-cold results; your measurement of sperm numbers donated by dogs during the 'tie' was, in all but seven of 25 attempts to measure, frustrated by the sampling tube being kinked or compressed, so that 1-ml samples could not be taken – and two of the 'good' ones showed no sperm donated.

There is a rule for *good practice* in the selection of your results: if you would throw out 'right' answers (according to your hypothesis) as readily as 'wrong' answers, for your selection reasons, your selection *is* good practice (even if there *are* no right answers in the ones you throw out – there usually will not be). If this rule is followed, your selection will not, in principle, bias your results. One other thing about selection of results is to think about them carefully before discarding them; they might be a different experiment; remember how penicillin was discovered. There may be better results in the results you discard (for now) than in the set you keep.

It is also tempting to think that because an experiment didn't work the way you wanted it to, that is, the results were against your prediction, it has failed for some reason, and the results should be discarded. However, it could be that a little careful analysis will show that there is something operating, but that it is not in the statistics that you are measuring. A good example is the demonstration that infection of sows with the roundworm *Ascaris suum* has an effect on their piglets' immune response to the same parasite[8]. The first experiment gave a very encouraging result: a difference between the two groups (litters of piglets with and without infected mothers), but, very surprisingly, the piglets with exposed mothers had a higher parasite burden than those without. Such a result required confirmation, and so the experiment was repeated, but much to everyone's chagrin (especially Jaap Boes, the student doing the work), the same result was not apparent. Oh dear, two experiments, one with a positive and interesting result, and the second failed to repeat it. At that point it looked like a chapter in the book of life experience rather than a thesis. However, it transpired, after some reanalysis by Shana Coates, another student in the group, that the results were completely consistent when considered from the viewpoint of changing the variance rather than the mean. Litters whose mothers were infected had a much lower degree of variation in their parasite burdens than litters whose mothers had never 'seen' *Ascaris*. Experiment saved. It is always worth spending several days reconsidering the data before discarding them as a failure.

There are, however, many unconscious biases that creep into experimental designs; the good scientist is conscious of more of them. We simply cannot check all of our assumptions, and some of them may be wrong: there may not be sodium chloride in the new bottle so labelled. But life is too short always to check *everything*. And the sodium chloride would not be the first thing to check if the experiment didn't work (gave a null result).

One common problem in chemistry and biology laboratories is the water supply. In several buildings in the Weizmann Institute in Israel, the general supply of specially pure lab water was labelled 'double-distilled water'. Because it didn't give 'clean' results in immunofluorescence work with spermatozoa, JC tasted it – awful! Tracing the supply showed resin-column-filtration, followed by distillation. This was 'comparable to whisky', he explained to colleagues. 'The resin takes out nearly all the ionic chemistry, but puts *in* a lot of biological chemistry. The bacteria living in the column secrete polysaccharides, small proteins and other (often complex, coloured) organics, especially if they are illuminated. This is like whisky manufacture, which uses pure spring water with few minerals, adds organics and then distils. Many of these organic products, or their breakdown products, pass the single distillation of both whisky and lab water.'

Much of the variability in blood-clotting properties of the early kidney-machine dialysis membranes was due to the washing in such

---

[8] Reported in Boes, J., Coates, S., Medley, G.F., Varady, M., Eriksen, L., Roepstorff, A. and Nansen, P. (1999) The role of maternal immunity in experimental *Ascaris suum* infections in young piglets. *Parasitology* **119**, 509–520.

water (the results differed between 'distilled' water left in the reservoir overnight, and freshly filtered water, which was better). In many other laboratories the water is labelled DW, for example, but is in fact produced by running tap water through – non-sterile – resin columns; these take out the chemistry – the meter reads 20 megohms or whatever – but they put *in* a lot of biology! Water supply is the classic 'Well, of course it's all right, everyone uses it!' problem[9].

Other not-thought-about biases depend upon our too-ready reliance on authority. Nearly always, it *is* sodium chloride. The mains electricity supply, however, in – most of – the UK is 230 V, not, as advertised, 240 V. In these days of microelectronically regulated power supplies, this is not serious – most power packs deliver exactly the right output from any input from 100–250 V. But many centrifuges made before about 1980 work from a variable transformer, and give the wrong speeds if the technician is casual – many human spermatozoa have been lost, being not firmly pelletted when washed, because they were spun too gently (speed is not linear with voltage). Some families, in consequence, have fewer children.

There is another, more direct way in which authority can bias. 'It's OK, I've washed the pipettes' makes one seem discourteous if one does them again. 'The alignment of that Nomarski microscope was set in the factory. *Don't* mess with it!' discourages junior workers from trying to improve it (objectives are set to about 1° at the factory; the polarisers and prisms can be set to give dramatically better results; 0.1° is attainable with simple tools and about two hours of patient adjustment). 'Have you tipped anything back into the sodium chloride bottle?' always elicits a 'No, of course not!' but many, apparently competent laboratory workers do tip back their surplus into the stock jar – with whatever was on the spatula or the balance pan.

There are other, even less obvious ways of biasing the results of an experiment or a series of observations. A postgraduate student was investigating the subtle differences of behaviour of female mice resulting from their uterine position as fetuses[10]. If they had been between two female fetuses (FF), they were rather passive females; if between two male fetuses (MM), aggressive and liable to overt male-type behaviour when adult; if between a male and a female (FM or MF), 'normal'. Because she worked with MMs, contrasting the behaviour with FMs, she mostly bred from the FF animals. These passive females reared their babies – all of them – to be very passive and unresponsive. After about four generations, the originally described aggressive behaviour of the MMs could not be elicited in the selected stock; the original strain was not available commercially, and the work almost foundered. This is a socially conserved trait which

[9] Incidentally, reverse osmosis (RO) systems for usable water are easier to control – but the water through the one RO membrane may well be more suitable for biological experiments than the 'polished' water, especially as the system is usually not assembled sterile. Of course, if the water is to be used as a solution of chemicals such as biological saline or bacterial media, the levels of inorganic impurities even in Analar chemicals makes the polishing pointless.

[10] Turnock, M.E (1993) 'The effect of stress and intrauterine position on reproductive function in the house mouse, *Mus musculus*'. University of Keele, UK, Ph.D. thesis.

crosses the generations, not genetic selection; but the research programme almost sank nevertheless.

Using laboratory-bred stocks guarantees that much of the behaviour (at least the breeding behaviour), physiology (responsiveness to temperature stress, for example) and especially immunology (acquired resistance/tolerance to common parasites and symbionts) is *not* like that of the wild organism. It does not take many generations of breeding in captivity to cause dramatic genetic changes. The blue gourami, *Trichogaster trichopterus*, the national fish of Singapore (it is on some postage stamps), is a food fish which grows to nearly 2 feet (60 cm), and rarely breeds below 45 cm (because it needs to find and defend a large territory). This species was imported into the USA and the UK in about 1924 for the hobby of keeping tropical fish, and has been inbred ('I'll have half a dozen of those juveniles please!') and selected for early breeding in grotesquely small tanks. In 1946 it grew to 8 inches and bred at 6 inches (JC bred them for cash); it now (1998) rarely grows bigger than 5 inches and breeds at 3. Is it the same species? It is convenient for home aquaria and for student projects, but arguing from the results to the wild fish is tricky. Yet this fish is still called, so authoritatively, *T. trichopterus*.

10.4 Authority and reductionism

There is another bias from authority, more subtle yet. Although your results now enable you to explain phenomenon X by your hypothesis (it's caused by A, but it's a good idea to have some B there too), all the rest of the causality in which your experiment is enmeshed is also affecting your results. If your supervisor is an authority on A and its effects, your hypothesis will be of the form A causes X. When your friend, who has been working with B in the next lab, does the same experiment to test her hypothesis that B causes X, you are amazed to have her use your work as confirmation. 'Sure,' she says, 'B causes X in the presence of an excess of A!' This can be sorted out over a drink by the four of you.

Some effects of content-versus-context, however, are more difficult. In the 1960s a UK radio programme 'Woman's Hour' was asked by a listener why hot water, in a metal ice-making tray, froze quicker than cold water in the ice-making compartment of her fridge. The presenter, having consulted a refrigerator manufacturer, denied that it did; hundreds of women rang in, confirming that it did. The programme asked refrigerator manufacturers to try it; they did: cold water froze first. Many more housewives rang in to say, 'Not in my fridge!' The BBC asked the engineer Eric Laithwaite for advice, and to try it. He did: in his home fridge, hot water froze quicker than cold – in trays next to each other! In the lab fridge, on the other hand, physics reigned and cold water froze first. Try to think what might have made the difference (remember that fridges didn't automatically defrost in those days). Eric rang JC, and we invented an explanation. In the home fridges, the layer of ice crystals stopped the metal ice-cube tray from contacting the conductive metal of the freezer, but the warm water melted it, allowing the tray to make better heat transfer. In the manufacturers' fridges, and in Eric's lab, there was very little ice. We have heard another suggestion too: warm water starts more vigorous convection currents, which continue to assist heat transfer, whereas cold water doesn't *start* convecting and so does not transfer heat as well. Mmmmm. Can you design an experiment to test this? Better, do we have information above with which it is incompatible?[11]

Let us assume (despite the manufacturers' evidence) that both mechanisms can operate. We now have two explanations for one phenomenon. Which do we choose? Popper gave us some guidelines here. First set up a list of hypotheses. We have two, that the anomalous rapid freezing of warm water is caused by convection or by contact with the cold metal. Others could be that it is caused by the warmth in the ice compartment switching on the control element, turning on the freezing mechanism. From the information given above, how do you decide?

Because the universe is not uni-causal (one cause for each event), X is not 'caused' by A, but by whichever variable (B, C, D...) is above or below some threshold; we choose to illuminate A's causality by making sure we have enough B, C and D, calling them parameters. This is a general problem with scientific observations and experiments, which must be related to the beliefs – prejudices, priors – we have about the general circumstances to which our hypothesis relates. Most (all?) experimental designs assume 'all else being equal' (*ceteris paribus*); only the 'internal differences' are investigated, because the context is the same for all 'plots' (in ANOVA, for example), all parts of the experiment. This directs our thoughts to reductionist, analytical hypotheses, and away from the general relations of the system we're interested in, away from its natural context.

This idea leads us into one of the great arguments in today's science, the question of *reductionism*. In its most naive form, this philosophy suggests that the phenomena of biology are 'really', or 'fundamentally', or 'basically', chemical in nature, so that an explanation of a biological event should *really* be given in chemical terms. This gives chemical explanations authority over biological questions. This is why biochemistry has such an honourable status among biologists; there is the strong consensus that a biochemical explanation is usually 'better', more satisfying. If we can say that the stickleback displays to rival males because *this* gene codes for the enzyme which results in

---

[11] This has been called the Mpemba effect, after an African student who discovered it independently; our referee suggested the URL www.physics.adelaide.edu.au/-dkoks/Faq/General/hot_water.html. See also Auerbach, D. (1995) Supercooling and the Mpemba effect: when hot water freezes quicker than cold. *Am. J. Physics* **63**, 882–885.

red colour on its throat, *this* one for the protein which synthesises the peptide which turns on aggression, and *this* one for the red-receptor pigment in the eye, then it seems nowadays to be a much more 'complete' and more 'scientific' explanation than those involving territory, game theory and evolutionary advantages. If we can talk about electron orbits in potassium salts, reducing the chemistry to physics, that also sounds as if we're getting closer to the 'real' causality. The climax idea of this philosophy is, of course, the *theory of everything*, which suggests that underlying all the causality we see is some deep structure which can be embodied in an equation linking all the physical phenomena into one idea. In the 1980s that looked to many scientists like the overall aim of science.

Then some thinkers began to realise that science did not converge downwards, with fewer and fewer ideas – quite the reverse[12]. When scientists looked into the gene for making red-sensitive pigment in the stickleback's eye, it turned out that the pigment is not the direct product of a gene, but a derivative of vitamin A that requires at least six enzymes for its synthesis, each from one of six different genes. To understand what makes the retinal cell different, so that it has the function of red reception, requires that we know about the different genes which are turned on in *that* cell, compared with the green-receptor cell, say. And it turns out that the cell is not *red* sensitive: as in our own eyes, which *seem* to be red, blue and yellow sensitive, the photoreceptive pigments overlap considerably in their reactivity. Therefore, the stickleback retina and optic tectum (the first visual area in the brain) have to do some computing to get 'red'. The neural connections in the retina and the connections up the optic nerve into the brain have been set up, during embryology and during the functioning of the system, differently for males and differently again in the breeding season. OK, it's complicated, with many genes interacting.

But that's not the major point. If we want to explain this appreciation of 'red' by sticklebacks, we need not just one authoritative physicist and one authoritative chemist, but perhaps ten different specialists to tell us about the absorption of the light, destruction of the photoreceptive molecule, the regeneration of these molecules, and the transduction into a nerve impulse; then we need a communications specialist to tell us why that train of nerve impulses rising in frequency, while these two others decline, tells the brain 'red'. We need more and more specialists as we go 'down' the trail to the physics, not less; the *reductionist nightmare*. Reduction of biology to chemistry and physics does not make life simpler, but more *complicated*; fields of expertise multiply. So advancement in scientific disciplines by publishing more papers, by word-paper miles, gives the biochemist a head start: every phenomenon in biology can generate many more papers on its chemistry than on its biology, and this may account for the pre-eminence of

---

[12] Cohen, J. and Stewart, I. (1994) *The Collapse of Chaos; Simple Laws in a Complex World.* Penguin, Viking, New York.

the biochemical disciplines. They *look* more complicated too, and have prestigious mathematical equations; they certainly don't augur well for one understandable equation at the bottom!

In just the same mistaken way, some people believe that 'under' the whole human species, engendering all of us, were an Adam and an Eve. Yet only a small bit of observation reveals *divergence* in your ancestry, not convergence: you have two parents, four grandparents, seven great-grandparents (probably a cousin marriage) and so on. Just as the Adam and Eve philosophy was promoted by the authority of the Church, so the reductionist philosophy is promoted by the authority of senior scientists (those who examine Ph.D. theses, for example).

10.5 Cycles of explanation and scales of organisation

We hope that we have convinced you that explanation of observations and experiments, and therefore our scientific understanding, is dependent on the definition's context and content. A simple phenomenon such as colour display by a small fish can be seen as 'explainable' from a large number of angles. But these tend to be directional: one can either go 'downwards' (to chemistry and physics) to describe something in terms of how it happens, or one can go 'upwards' (into more biology) to describe something in terms of why it happens. But, in biology at least, and probably everything else, these directions lie on a circle.

Figure 10.1 shows this circle. We now know much about DNA – we are publishing genomes at a great rate. Hurrah for modern biology. Now attention has switched to the transcriptome – which genes are being transcribed (are 'switched on'), when and under what control? The proteins created by translation of the RNA are themselves a complex mass of (non-linear) interaction: many proteins require other proteins (chaperones) to help them fold up correctly so that they can work. Translation and transcription themselves require proteins. So all the arrows in the figure are bidirectional – the influence extends in both directions. In fact, they probably ought to be subcycles. Each level of organisation creates the context for its own content – thus cells use their proteins to create their proteins; tissues provide the chemical cues for cells to differentiate appropriately to create the tissue. All, that is, except one. Populations influence DNA (through evolution), but DNA's influence on populations has to be manifest through the individual (via biochemical, cellular and physiological pathways).

In terms of explanation, the answers to the how questions (the mechanistic explanations) are found by going anticlockwise; by

Figure 10.1

The biology cycle.

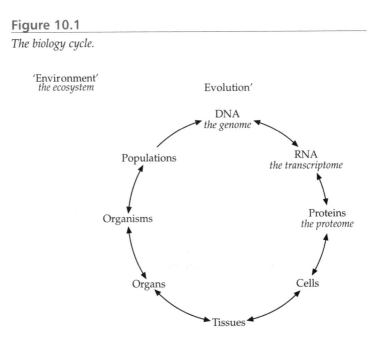

referring to the next level of organisation down. Thus, physiology is explained in terms of organ function, which might be explained by tissue and cellular level phenomena, which in turn have explanations in proteins, and ultimately DNA. But DNA is not the starting point. It is itself determined by evolutionary processes that are manifested through competition within populations (in the context of specific environments), where the individuals with the best physiology are fittest. Thus, these explanations are circular, and provide us with a deeper understanding of the processes that support the next level up. The reductionist nightmare happens when we step off the cycle, and delve in detail into the mechanism in any one of the steps.

The why questions (the functional explanations) are found by going clockwise. We have the DNA we have in order to create the RNA, proteins, cells, tissues, etc., that we need to create a human, which has to survive in a population of humans (and its environment). But the cycle has to be completed with a Just So Story: we have the DNA that we have because we have that DNA. Why do we have the genome that we do? Answer 1 (clockwise): because human evolution has produced it that way. Answer 2 (anticlockwise): because we need the proteins that it creates (if it was any different we wouldn't be human). Answer 1 is clearly more satisfactory: it is an explanation that explains to some degree.

This is a bit imprecise, but there is a real example that expresses just how inter-linked these levels are. One of the heat shock proteins (HSP90) was thought to be just a chaperonin like all the others: when an individual is stressed by, say, temperature change, these proteins help other proteins stabilise and retain their shapes. But it turns out that HSP90 does more than this, with far-reaching evolutionary implications. When proteins mutate, they would, if left to themselves, adopt different shapes. But HSP90 seems to constrain them into the ancestral shape even when they have several mutations, so it hides these mutations from phenotypic expression, and has been called an 'evolutionary capacitor'. Like an electrical capacitor that stores electricity, the mutations are stored and only released and expressed to become fuel for evolution, when HSP90 is all being used to deal with heat or other stress. When an individual is stressed, then, hidden variability is exposed when evolution can best use it. This example shows that the cycle we have used for explanation is more than just a hand-waving exercise, the functions of HSP90 can be seen only in context against this cycle.

10.6 Better explanations

Even if the reductionist nightmare were not a good reason for disbelieving naive authority reductionism, there is another problem with reading causality 'up' or 'down' the system: the 'Krebs' cycle is *the* chemical cause of that cow eating that grass. This is that higher-level phenomena are not limited in their causation to one set of causes underlying them, they are not uni-causal (you should have A, B, C causing X in mind).

Think of a bridge, linking an island to the mainland. This can be made of rope, steel or concrete – or even of nothing (and we call it a tunnel). It *cannot* be the case, then, that the bridge property or bridgeness is uniquely caused by what the bridge is made of. Similarly, the stickle-back probably has several alternative possibilities for its red pigment, or for the hormones which turn on aggression. Whichever is used, the stickleback male, in season, is aggressive towards other red-throated males. The underlying causes, the mechanisms, may be replaceable with others. This idea of 'replaceability to all intents and purposes' is called *fungibility* by lawyers, and it is a useful concept for the philos-ophy of experimental science. If we agree that there may be other kinds of animals using red colour as a territorial marker like the stick-leback – there may even be extraterrestrial life forms using red as a territorial marker, but based on non-DNA genetics – the DNA (parochial) explanation for why the stickleback – our stickleback – behaves thus becomes less satisfying. We want some general statement about organisms using colour to communicate. It doesn't matter what the bridge is made of, for many purposes, but what and how, perhaps even why, it joins.

If we don't 'explain' biology by statements about the chemistry, or explain physics by mathematical equations, how *should* we seek explanations? This is an important question, because what we do with the sticklebacks to see why their chins are red depends on how we answer. If the answer we seek is the reductionist one, we would have to start by some equivalent of taking out the red colour and analysing it, which probably entails some version of putting the fish in a food blender and then extracting the substance we want; we can chase the genes in other fractions of the soup we get.

If we intend to explain the colour by metaphor and class: then we need to see stickleback colour in specific contexts, such as the following:

- If it is a *territorial* colour, we can set up sticklebacks in territories and stimulate them with coloured models.

- If it is a result of seasonal *hormone* changes, we might affect the hormones by day length, by injecting other hormones or by dissolving them in the water.

- If it is the behavioural *ranking* of males, we can set up aquarium tanks and see whether male rank relates to amount of colour, assessing male rank by female choice, or by staging fights between the males.

- If it is …

The list is, of course, endless: any property can be seen in a vast number of contexts. Our intention is to make people say, 'It's another one of *those*!', when we explain how our observations or experimental results add to our information about the red colour – 'It's just like the robin in my garden!' This is usually best done by *eliminating* possibilities rather than adding to an ever-increasing list of causes or mechanisms. The best explanations are comparative.

The experienced scientist tries several contexts, and sees which gives an interesting answer – territory or social ranking? Or she comes from the other way: I want to ask such-and-such a question about territoriality – should I use sticklebacks or robins, feral cats or wild wolves?

But the object, again, is not to explain the observations by chemistry – *content* – but to put the endocrine, seasonal, food- or mate-related behaviour into territorial *context*.

10.7 Causation and causality

There is even more confusion about causality in scientific explanations than about prediction and testing. It is at least unwise, then, to assume that events have but one cause: they are the result of things (everything?) that went before, and/or result from the coming together of several disparate, perhaps contingent events. This is reflected, in the scientific world, by *analysis of variance* kinds of questions which we might call 'multiple observation – or undesigned – experiments': what factors, and how much of each, contribute to this result?

A lot of the confusion is due to the several ways that we use the verb 'causes'. Gravity *causes* things to fall. If there were no gravity, things would not fall. This is the way in which people mistakenly take the word to be meant throughout science. Smoking *causes* lung cancer – what we mean is that smoking increases the risk to individuals of developing lung cancer. (Gravity does not *increase* the risk of things falling!) We would still see lung cancer even if nobody smoked, as there are other causes. Radiation also causes lung cancer. In fact, somebody exposed to radiation may be *more* likely to develop cancer than somebody smoking – the risk to the individual is greater with radiation. At an individual level, radiation is a greater 'cause' of lung cancer than smoking. But more people smoke than are exposed to radiation, so that at a population level smoking is a greater 'cause' than radiation.

Perhaps there is something wider here – the interaction between population and individual is in some ways analogous to the relationship between mean fields and the things in the fields. Epidemiology is largely about this relationship between relative risks and population-attributable risks, which is analogous to individual behaviour and population behaviour.

10.8 Hypothesis, paradigm and progression

Reread the sperm-competition experiment in Section 9.7. The philosopher of science Popper has already been invoked; he would have approved the sperm-competition experiment, because the very unlikely result was anticipated, and obtained. Another philosopher of science, Thomas Kuhn[13], might have seen this experiment as demonstrating a change of *paradigm*, of the way in which biologists thought of sperm populations. He might have seen this as a move from homogeneous populations to heterogeneous populations. He would be concerned to discover *how many* biologists changed their minds upon reading the results (not many, says JC; perhaps a third of reproductive biologists, 30 years on). But a much less pretentious test of whether something is successful science has been proposed by another philosopher, Lakatos[14]. He asks whether the research programme is *progressive* or *degenerate*. Progressive programmes continue to generate new experiments, which provide *interesting* (potentially mind-changing) results. In contrast, degenerate programmes run out of new and interesting ideas, and the scientists concerned are forced to reiterate the same story at a succession of meetings.

The sperm-discrimination programme was manifestly progressive: it produced a variety of clinically important conclusions, widely accepted but not close to the original experiments. The hypothesis was refined to two sperm populations, discriminated by antibody in the female tract (Cohen and Tyler 1980)[15]. A major criticism was that there was not enough measurable antibody in normal (uninfected) rabbit or human tracts. But when we looked at sexually excited tracts, we opened a whole new field of research: not only was ample antibody secreted, but the immigration of white blood cells into the female tract was most dramatic[16]. It explained many anomalous clinical cervical smear results, which looked pathological (many leucocytes) but were from healthy-looking tracts. Women were usually asked to return for another smear, at great cost to the smearing system and anxiety to the women. If we were right, the presence of spermatozoa made the presence of leucocytes physiological (normal), not pathological. We designed and performed several series of observations of cervical smears of artificially inseminated women (and rabbits); we also experimented with some other components of sexual behaviour (such as semen from vasectomised men, which did not elicit leucocytes). This new work amply fulfilled Lakatos' criteria of a progressive programme, and many laboratories are now following this up[17]. Further, clinical cervical smear programmes worldwide now save money, and anxiety, by not recalling women with leucocytosis if they have sperm too.

---

[13] Kuhn, T.S. (1970) *The Structure of Scientific Revolutions*. University of Chicago Press, London.

[14] Lakatos, I. (1974) Falsification and the methodology of scientific research programmes. In *Criticism and the Growth of Knowledge* (ed. I. Lakatos and A. Musgrave), pp. 91–196. Cambridge University Press.

[15] Cohen, J. and Tyler, K.R. (1980) Sperm populations in the female genital tract of the rabbit. *J. Reprod. Fertil.* **60**, 213–218.

[16] Smallcombe, A. and Tyler, K.R. (1980) Semen-elicited accumulation of antibodies and leucocytes in the rabbit female tract. *Experimentia* **36**, 88–89; Pandya, I. and Cohen, J. (1985) The leucocyte reaction of human uterine cervix to spermatozoa. *Fertil. Steril.* **43**, 417–421.

[17] Thompson, L.A., Barrett, C.L.R., Bolton, A.E. and Cooke, I.D. (1990) The leucocyte reaction of the human uterine cervix. *J. Reprod. Fertil.*, Abstracts **5**, 12.

[18] And you are not allowed to say, 'So that they can fly', unless the child can be guaranteed to ask, 'Why?'

10.9 Conclusions

The main conclusion is that explanations rarely explain. They can't. Explanations do not exist (in science at least). If you think that this is a gross, and wrong, generalisation, just try to explain to a 5-year-old why birds have wings[18]. Explanations are stories with different degrees of believability. Better explanations help us to understand better and have wider applicability. Science itself is just a collection of the most satisfying stories around at the moment. The true goal is to understand, and this is usually achieved by having several simultaneous 'explanations' that don't quite fit together.

Science progresses through discoveries of things that weren't known (e.g. the existence of DNA, chaperone proteins or natural selection), through the better understanding of those things (e.g. the structure of DNA and its methods of replication or HSP90 or inclusive fitness), and through the generation of general ideas for the way that things interact (e.g. the idea of the founder effect in populations, which results in populations with much reduced variability in DNA). In each case, the satisfying explanation has a beginning (an observation), a middle (a method and results) and an end (therefore…). This should be comforting – you are seeking to provide explanations with this structure – stories to tell that everybody likes, but that nobody has heard yet. In some of your listeners, these stories will mature into understanding, and you will have added to the sum of human wisdom.

Content
and context

11.1 The postgraduate in context

This chapter was originally intended to give advice on how to get
your work into the right context, with wider applications. We were
going to advise that you read popular books and magazines on your
subject, and visit other laboratories and research teams. How do the
Internet encyclopaedias describe your area? What is the government's
view on the developing technology? This is particularly important if
are doing something of high public interest, perhaps in medicine,
criminology or genetic manipulation.

However, the first drafts for this second edition carried us into a
consideration of those of our students who had had to take some time
off, and had found it necessary to see themselves *and* their work in a
wider context. We thought we had some valuable things to say
because so many students go through a scientific 'identity crisis'.

If you have read as far as this, you're almost certainly well into your
research project, probably in the second year at least. This is when

most postgraduate students begin to get depressed about their projects, and start to feel as though they are swimming through thick lentil soup. This is when their need for books like this is greatest. But the help they seek is often very specific indeed – too specific to provide the overview that is really needed. We can't give the specific advice on your project, but we can give an overview.

Here are some usual, and then some less usual, problems; all of these have been experienced, and survived, by our postgraduate students. This is not a complete list of all problems ever experienced by students. Nor do we think that our students have had an especially difficult time[1].

11.2 A list of real difficulties you might face

1. You find that what you thought was complete understanding between you and your supervisor has turned to mutual incomprehension – and not only about your project. You avoid each other and/or you are sure your supervisor is stupid, inadequately informed or having an affair with another student.

2. You have found out enough about the academic geography of your subject area that you are afraid that the gang in Chicago will publish long before your results are ready. Or one of ten other teams.

3. You have found out that your supervisor and so your stance are very heterodox, and you're worried that you won't get the 'right' results at all. All the published work in your area contradicts the position you have espoused. Other workers in the field and staff in your department are pessimistic.

4. You have found that the apparatus you're using or your research material is very old-fashioned, so that you might not be employable by groups that use the nice, modern shiny machines/kits. Is it worth going on?

5. You have decided that making music or artistic sex, or having a close loving relationship, or learning Cantonese or trampolining is (or possibly all of them are) much more interesting and taking much more of your time and attention than your research topic.

6. You have become subject to migraine attacks or other headaches, to falling asleep in the library (or when working

---

[1] Although we also hope that it wasn't easy for them.

through references on your computer), or to loud and painful indigestion. Or you have discovered that a relative/lover has MS or TB or HIV, so you can't engage your mind in your research.

7. You have discovered that your mathematical, statistical or essay-writing foundations are totally inadequate for what's expected in your research group. You are constantly feeling embarrassed by your peers.

8. The experiments you did in your first year are not possible any more, because money has run out, the apparatus has broken, another student has taken over the source of organisms or you can't go to the seashore to collect because there has been an oil spill.

9. You are realising that all the others in your research group are much brighter, more competent and more socially engaged in the team than you are; and there isn't a place for a loner/incompetent. Or you thought your facility with the new language was adequate, and it just isn't.

10. You get pregnant or your partner does, or you go bankrupt, or your parents part and one of them needs you for company, or you are caught in a criminal activity and are to spend months or perhaps years in prison.

All those have been successfully surmounted by postgraduate students in our own direct experience. On a few occasions the decision was to leave research. But, if you want to stay, it is usually possibly to take some time off, keep reading around the subject or keep in touch via computer, and re-engage after weeks, months or even a couple of years. It can be useful to spend weeks or months at another establishment, or to change research material while addressing the same problem. Or, indeed, to change the questions you are asking and go on using the same material. Read on, because all of these issues become resolvable into a research-compatible format if you can see your own research in context.

11.3 Honesty and dishonesty

The real problems are not falsification of data or results. The real problem is misunderstanding what science is about, and inventing false worlds, false results and a false universe. The most common problems are those people who foment misunderstanding, often by requiring others to hold to absurd standards: 'You promised you'd read Smith and Brown over the weekend and explain it to me on Monday, so we could repeat their work on Wednesday in time for the Friday meeting!' This forces the pressured student (once it was the supervisor pressured by the student) to assemble a set of kind-of results.

Rarely the falsely constructed universe is just on paper – it's more likely to be in the mind, a stance committed to because it's been discussed with others. There are probably two ways in which this dishonesty shows itself. First, and less common, students make up some results. Good supervisors, in our experience, are those who – when they find out the results are faked (and we don't know how often we don't find out) – use the figures in a 'what if...?' exercise. 'If the world really is like that...' allows us to show what a pre-experiment statistical exercise does, and relocates the student on a productive path without too much embarrassment. After all, doing 'what if...?' exercises are exactly the first step in experimental design. And inventing a possible set of results is just what we have encouraged you to do, so that you understand the real ones better when you get them. It always is a temptation to intercalate them into the real sets, we suppose; and we don't know how often it's done. There are ways that expose made-up, as distinct from 'naturally generated', lists of results. These methods usually rely on looking at variability – it is far more difficult to fabricate natural variances than it is to fabricate means. In fact, it's usually much more trouble to construct Zipf-proof[2] lists than it is to do the actual experiments or the observations. Of course, students are as likely to be given falsified data as they are to create it. If a collaborator gives you data, how do you work out if they made it up? This is a worthwhile line of thought, because it will sometimes reveal ways that you can develop novel analytical approaches and understand the data better.

The second, and far more common, fabrication is hiding a set of results that don't fit prejudices, or making assumptions that are blatantly untrue but which the student would like to be true. In this case, the work wanders away from the issues because the student takes a path away from confrontation with the problem at hand. Instead of making up some figures, which can be a useful exercise – or used as such – the student falsifies his position in respect of other workers, especially colleagues in the same team. This isn't science because you are believing what you want to. 'I decided to use B instead of A, because the A apparatus is always in use by X, who bullies me about it.' Or 'I've tried and tried to get the M process to work, and even though I've been shown ten times it doesn't work for me, so I've done P for the last six months and my results don't make any sense!' Occasionally students are seduced away from the problem at hand by a very attractive computer programme or a shiny new

---

[2] Look up Zipf's law and Benford's law on the Web.

biochemical kit or a spectacular microscopic picture – they work very hard at the new procedure and hardly think at all.

The supervisor's job is to guide you gently back to the data. The careful handling is required, because the aesthetics might engender a whole set of new questions: JC had a Pakistani who saw zebra fish embryos and was caught – and after a honeymoon period (and some shall-we-say 'cryptic' results) produced wholly new questions. On the other hand, a student 'caught' unproductively at a kit, a microscope or a new piece of equipment is keeping others from using it more productively. We've succeeded in engendering productive thinking by requiring the postgraduate student to teach an undergraduate project on the topic.

11.4 What is postgraduate research for?

If there are so many painful ways for postgraduate research to go wrong, what are the reasons for attempting to mend the situation and get the research done? Why not simply go out into the world with your first degree and get a 'real' job?

Firstly, if you're committed to science and research, there will be no question: this is your way up, the way to be trained as a scientist. Move to another lab or another team, perhaps. But leave the profession? – no.

More pragmatically, with the UK and other governments attempting to get half of the young population educated to university level (this is not the place to discuss how mad that is), your chances of finding a good, interesting and well-paid job are small. And this doesn't take into account whatever it was that made your postgraduate research programme go wrong.

Thirdly, and this is the real issue: if you are bright enough to get trained as a scientist, and what has gone wrong can be remedied, then you will be improved by the experience of mending it. There are few Ph.D. candidates who have had a smooth ride; part of the training is to deal with bends, even hairpins, in the path. And it really is so, that those trials which don't kill us make us stronger.

So let us look at the context as well as the content of your research project. There may be eight of you involved, say, two postdoctoral fellows, perhaps three postgraduate students and three technical support officers (what we used to call technicians, secretaries or personal assistants). You are addressing a question invented by an academic that you see sometimes. She put in for a grant four years ago, and it was awarded, somewhat reduced, two years ago. It pays the eight salaries, and consumables, and has provided equipment, and will run for the three years during which you hope to do your Ph.D. project. Perhaps you'll be thinking of writing it up after you've left the post here. (This is becoming much more common: it used to be the case, and was for all but two of JC's full-time postgraduate students, that it was all to be done within the three years.) There's a group in Edinburgh on much the same topic, and one in Pittsburgh; there's a group in Chile, too – you don't know much about them – and probably some workers in Japan (you know one is at Yale for a year). You're in your second year, and for one of the reasons above (pregnancy, trampolining, prison, or the world crochet championships) you want time out; perhaps six months, better a year. But you can't talk to boss lady; she seems not to listen to you. You've talked to the PA, who wasn't helpful: she said what you knew already, that your boss lady is very concerned to get the questions in her grant proposal addressed in the time she promised. You would feel very bad about leaving at this time. But you know a new postgraduate who wants to join a research team, or you can take some time to join the Pittsburgh lot (it's the crochet centre of the world), or your partner hasn't a job and can look after the baby … or … or … or …

You are worried about what's the best thing to do, in all the circumstances. You think that you must make a judgement, taking into account all these different weightings. You get headaches worrying. Of course you do – you're a responsible person, you don't want to 'let other people down'. Here we can help. You are distressed because you believe that a judgement about what's best is on your shoulders. It isn't. When you were appointed/offered the studentship all these possibilities were there, and the experienced people administering it, including the boss lady managing the programme, knew that something like this might happen. What they want from you is not the complex judgement – that's *their* job, or the PA's. What *you* must do is make a decision. Choose among the options. Perhaps ask the new postgraduate if he'd take the job, informally. But then commit yourself to a course of action, and tell your colleagues what you've decided. It is entirely possible that they'll give you another, better option that you've not thought of, or thought wasn't possible. In any event they will be happier that you have decided and that you're not moping about, holding up work.

The judgement-versus-decision point is equally valid if you're just one postgraduate student, with one supervisor. Make your decision and go to him.

Making a decision commits you to that course of action in the unforeseeable future. Decisions don't depend very much on weighing up past situations. You must decide if your future will allow you to come back to research, even perhaps to the same research topic. If so, ask that a 'hold' be put on that, so far as is possible, and commit yourself to a time of return – and arrange to keep up with the literature. This is much easier, now that the Web is available from anywhere. Some universities can arrange that you can ring in to their library, and access professional journals from there. If you can't do that, you may have to subscribe to a professional journal, which will usually allow you access to the journal's website, and from there you can access other journals in the field.

Making a decision commits you to something more: it means that you must never say – or feel – that you chose the wrong option. You can regret a judgement, if you feel that you didn't put the right items on the balance pans. But when you make a *decision*, you should commit yourself not to regret it. Decisions should be 'future-proof', so that even when you look back in hindsight and know things that you didn't know then, you can reflect that, at the time, the decision was correctly motivated. That's a good principle for life.

11.5 Content and context

We hope you have read the above, thinking 'Thank goodness none of those things have happened to me!' That's a good reason for our making the list; it shows you all the things that might have happened that you have – so far – escaped.

But it has also, we hope, forced you to think, 'What if I *do* give up now? What difference will it make to me? To my parents, partner, children or spouse? To my colleagues? To my supervisor? To those who've given us the grant? To the area of work? To the planet?'

It might have encouraged you to see the work you're doing in context. You may be looking at the effects of a possible pharmaceutical, a herbicide, a new magnetic alloy, a new mathematical technique for a quantum-physics application, sperm from a group of infertile patients or a new diagnostic kit for an imminent epidemic. All of these have different time-scales: some of them have urgent but not imminent usage; some, like the diagnostic kit, may be imminent but not very urgent (because there are other kits that do the job). And so on. The cancellation or delay of most research projects will have the greatest impact on the student (you). They will also impact on your supervisor's reputation for getting questions addressed, papers published and grant contracts fulfilled. A few research projects have very wide implications, but in that case you'll almost certainly know of other able people addressing the same issues. If you really are the only one addressing a critical question, or believe you are, talk it over with other academics as well as your supervisor. Perhaps you aren't.

This exercise, seeing your work in context – even if you've only done it because of a threat to its continuation – might turn out to be the most useful preparation for the introduction to your thesis, and for your *viva voce* examination. Certainly the majority of theses don't have this contextual vision that improves the choice of content enormously. So stand back from your subject, and see what you're doing compared to colleagues, to academics, to managers, and to your parents and your friends.

Very nearly all postgraduate students anticipate that their conclusion will be that the work is trivial – and then they find that it is not, that it has a real place in the knowledge base of science.

But also read our postamble (Chapter 13): most often, the most important product of your work is not the advance in knowledge enshrined in your thesis, or the papers you gave at meetings or even published, on the way. It's the change in you. And that is most often helped by surmounting obstacles.

And you should have realised that this process does not stop with the award of a Ph.D. degree. Everything in this chapter applies to students of research, research scientists, regardless of qualification or age. The process of self-evaluation of thought and learning should never cease; the search for the boundary between content and context is actually a large part of what research is[3].

---

[3] The Revd Thomas Bayes clearly knew this.

Notes for postgraduate students

12

12.1 Where and with whom?

Whilst writing this book, we have largely drawn on experience with postgraduate students that we have supervised or seen supervised. Although this book is not meant to be a manual of how to sort out your life, but, rather, about how do research, we felt that a few words to aspiring postgraduates might be appropriate. Of these, this first section is the most important.

Be careful when selecting a research group to join. Firstly, ask yourself what you have to 'sell'. You have presumably got (or are about to get, if you're going straight into postgraduate research) a good first degree. We think it's very reasonable to go for academic research only if your first degree was well within your capabilities; that is to say, you got/will get a good first degree. In the UK system, that means a first or a good upper second-class honours degree (at a university or college with high standards). Some time out of academia, doing real things, is a plus; a smaller plus is having gone around the world on other people's money. You will have had some experience of an undergraduate 'research' project or two, when you were faced with a problem and, with varying and different amounts of help, you got on top of it. If you failed at it, but it really wasn't your fault, go back and read Section 3.2 on 'luck'. If it *really* wasn't your fault, read on. Be prepared to talk about this experience, as it may be the only useful conversational selling point you have. Find out what's happened in that field *since* you did your project.

Find out the social geography of the places you *could* go. (Your local 'career adviser' may not be a good source of advice – think what job he got, and whether he does it as if he chose it.) Do you want to be a member of a large group, doing research that is safe and boring, not initiated or managed by you personally? Or do you want to be a moderate-size fish in a little pond, your supervisor the one person in that department interested in your field, and you the only postgraduate student this year, largely left to your own devices? Think to yourself about the coming interview: who's buying and who's selling here? Am I pleading to be taken on, or am I much better than the average she'll attract so that I can ask about lodgings and so on, straight away? Do the 'What if she asks…?' routine on your way so you have your answers ready – or not, if you're better at unrehearsed conversations; many people are.

Don't only think about the money. As anyone with enough money will tell you, it's not important. And that's true: it's like air, or love: if you've got it, it doesn't worry you – it's not important. But *absence* of any of the three is desperate; this is not a symmetrical situation. You must have a set-up where your thinking time is not compromised, though. Some people really can do their reading and thinking in all the chinks in the organised day, on buses and trains, while eating and in the toilet. Others need to have hours of quiet; here the library can provide if your lodgings can't.

Staying with your family, unless they're academics who understand, is nearly always a bad idea. Most people won't allow you to think about something for hours, let alone days: it looks to them as if you're doing nothing, remember. Your mother will send you shopping. Home is cheap, but may be impossible for your science. Or not. Working in a bar at night is very popular with postgraduates; but do remember that your brain is your primary tool – don't abuse it. Think of yourself as an aspiring Olympic athlete, but brain and mind rather than body. Exercise your mind (Chapter 1) but don't forget your body, either. Take regular exercise, and we recommend swimming rather than rugby – not, unless you're very good indeed, boxing.

If you come from a different country or culture, practise contact with research workers of the culture you want to join. Contact someone you have heard of, or found on the Web, by e-mail or telephone, and ask if you might come in to his unit and talk to the workers there. Don't be diffident about saying, 'We don't do things that way at home' or 'My undergrad tutor told me this might happen, but I never thought I'd see it!' There are many, many ways of doing science. They are not all effective, successful or progressive, but there are many very different approaches. Successful managers of research groups know this, and will value your differences, your individuality, particularly that you come from a different culture with differently useful ways of thinking. (Of the eight wholly new ideas JC claims to have initiated, two were developed with Nigerian Ibos, one with a Hindu midwife, one with a Bangladeshi Moslem, two solo, and two with UK postgraduates.)

A few words here that you may believe to be not quite politically correct. We are not singling out races or culture for opprobrium here. Our experience has been that there are some attitudes that *don't* address scientific research properly; these are nearly all committed to the idea that there is a right path, independent of what experience tells you. They are fundamentalist and/or authoritarian. A geneticist colleague of JC's (a good Christian), lecturing to final-year biology students about the rate of evolution (Gould/Lewontin 'punctuation' had hit the headlines the previous year), said that students could take this on board 'without problems with their Darwinism. After all, Darwin had himself said in the sixth edition of *The Origin of Species* that he didn't know how rapidly evolutionary change occurred'. JC, in the audience because he chaired the evolution course, leapt up and stopped the lecture at this point. 'Even Darwin is not to be an authority! We know much more now. Darwin was clever, even wise – but he couldn't know about later discoveries.' And the students could disagree with Darwin if they thought the evidence warranted it, and would get higher marks from JC if they did! JC's Bangladeshi postgraduate student was in the audience too, a devout Moslem – and cheered him on. We have, however, met (especially but not uniquely) Islamic students who needed the results to conform to the textbook, and who suppressed those results that didn't. One of these, with JC, found to her surprise (and later advancement, academically and, we're sure, spiritually) that it was those maverick results that formed the basis of her Ph.D. thesis and her later eminence in her field. Teachers in the Indian subcontinent, particularly (but in some central European cultures too) seem, during the early education of the postgraduates we get, to imbue a sense of 'Your professor is always right' (perhaps they have retained the worst elements of British Imperial teaching – sorry). We don't like this. If we're wrong (as we often are, not least because we do imaginative research), we want to know as soon as possible, not to waste more time on useless ideas. Postgraduates who show us we're wrong (and most have, of course) gain our gratitude; perhaps a little grudging for a year or two, because we *are* human! Now that Chinese culture is contributing so much to science all over the world, we anticipate that new directions will be found again, just as when Japanese scientists became part of our world.

Every interview you have will tell you useful things about the field you're entering; listen carefully, especially to the postgraduate students who'll be showing you around. Ask awkward questions about past students during visits: 'Where are they now?'; 'What did they do?' Most importantly, talk to the current students and find out if they are happy. Or frustrated. Or productively unhappy – 'My experiments aren't working yet – come back next month' is a very positive sign. Try to see the group you're joining as a process, like a whirlpool or a fountain: water (i.e. people) is always changing, but the integrity of the group should remain. It's very positive when they can tell you of previous postgraduates now running their own groups.

12.2 More social science

Now you have been taken on to a research group, or you have met your supervisor to begin years of hopefully productive research. The biggest problem facing the newly appointed postgraduate is the switch from absorbing (consuming) information to producing it. From the age of 5 years, most people (if they are lucky) get *told* things at school. As we have said before, these are often 'lies to children' – stories which mesh together and explain everything. All of a sudden in their early adulthood, these same people are expected to try and disprove these stories, and to be critical of what is in the textbooks. It isn't easy, and many fail to make the transition. It comes as a shock to find out that nobody *really* believes the things said in undergraduate lectures that had to be regurgitated to get a degree. It can be even more of a shock when you are expected to rewrite such a lecture, and give it yourself to those innocent-looking undergraduate faces. Then you find how tempting it is to gloss over complications by claiming things to be known, understood and demonstrated, and how hard it is to persuade these aspiring scientists that science is not truth, but just the best defence we have against believing what we want to.

Involving yourself in research is the first step (it doesn't particularly matter what you do). The crucial thing is to be able to interact with fellow researchers, especially other postgraduates and postdoctoral fellows. And interact *about science.* Firstly and most importantly, good research groups are relaxed about discussing the work of the group informally. A good supervisor is one you can go to and say, 'Petra, why did I get this odd point among these good results?' and have her say 'That chromatography apparatus was last used by Victor, you know – and he doesn't clean up as well as the rest of us' or 'Is that what the system *should* do – and all the other points are artificially clustered?'

There are different informal styles in different countries, but there should be easy communication and criticism of *all* the groups' experimental results and difficulties across and up and down the hierarchy. Bad habits, like appending names to papers whether or not the author has been involved, appear and become part of the mythology of the group. Celebrating each acceptance of a paper (by a refereed journal) by a communal Chinese meal, or a lab party with everyone bringing in food and drink, seems much healthier. Other healthy signs (to our rather personal prejudices) are a communal eating habit for variable parts of the group, with no taboos about talking work, quite the reverse! Lunchtime and teatime are often the best surroundings for sorting out experimental problems, but there should also be a special place set aside for thinking; you should be pleased to see one of the team stretched out on the sofa, not concerned that she 'should be working'. In our experience, that's when the thinking is done. If there is a taboo about 'talking work' at mealtime or after work, what we call a nine-to-five mentality, we would find that a very difficult life to lead. Most of our best ideas – and our most potent criticisms – have come with colleagues and students over coffee or a pint, or on the way to a field trip or a concert. We believe that science is a full-time activity.

The talking within the group is very difficult, sometimes, for the new postgraduate to accept. Science is, contrary to the received wisdom, a social activity. Those groups that get the social aspects of the science 'right', like those commercial enterprises which have their ethos proper to their function, succeed[1]. But the social aspects of a well-run, happy and happily critical scientific group are very different from the environment that the new recruit came from. That 'I don't know' or 'Does anyone here know about...?' earn just as many points as the didactic 'Yes, I read about that in last week's *Nature!*' is not easy for a new recruit from a less-than-relaxed university department. Particularly difficult for some cultural backgrounds, in our experience, is humour about the serious matters we're all concerned with. There's a particular kind of irony that British academia fosters that is very annoying to European visitors, particularly senior Danish and German academics, we have found, who are branded as 'having no sense of humour'. Actually, Irish and Hungarians (and occasional Canadians) do it better than the British. Those not in sympathy with this attitude say, 'You're not taking your experimental problems seriously', because we joke about them. We *are* taking them seriously, so seriously that we have to joke about them or weep. We're not taking them solemnly, with a long face; we think it's a social sin of the worst kind to be solemn about personal or 'family' problems. Only politics or religion deserves this depressing treatment, because they can't improve. A ban on talking religion or politics over coffee, now...

A pile of recent journals is frequently more useful than a well-organised rack, although the latter should be somewhere close; a rack

---

[1] Peters, T.J. and Waterman, R.H. (1982) *In Search of Excellence*. Harper and Row, London.

of recent publications by the group is also very useful, and gives the new recruit a chance to discuss the various things the group has done recently. It's a good way to find out who does what, too. However, it is often difficult for that new recruit to realise that the apparently least-productive member of the group (measured by word-paper miles) may be a very necessary administrator, catalyst, information source or goad to higher thinking!

It is not enough to do a good experiment, and get good results. Other people must hear about it, and their criticisms must be available for discussion as part of the incorporation of those results into 'science'. For the standard scientific paper, the editor of the journal uses a panel of respected scientists as referees; there *are* problems with this system, but in general it works well. Criticism of scientific thinking by one's colleagues is the most immediate, most proximate and usually most useful way of refining scientific thought. Take their time for *your* problems, and be prepared to give honest, considered criticism to them in return. One step more formal is the *research presentation*.

12.3 Giving an informal research presentation

There are two key things to remember. First, that there is absolutely no point whatsoever in doing research if you are not going to tell other people about it. Second, you know far more about what you did than anybody else does. The first should provide you with the impetus to overcome nerves; the second should give you confidence. Nevertheless, most people find presenting their work very difficult at first.

This is the kind of thing we have all done, from presenting under-graduate projects to one's peers – and one's examiners – to talking about one's own research as part of the assessment procedure for a new job. It is not the niggling, difficult, 'Why can't I get this to work when even that idiot in Pittsburgh can do it?' problems that need exposure here. You need a broad-brush approach. Typically, the audience are not specifically interested in your story; they are more concerned that you don't overrun your time than that your science is impeccable. These may be 10-minute or 60-minute talks, with 5 or 30 minutes' discussion, and the techniques are very different.

For a brief talk, take enough time (perhaps as much as 4 minutes) to make sure that nearly all of the audience know what question you're addressing, and how they'll have to change their minds about your subject if you get answer A instead of B. Then briefly tell them what you did (never mind the make of the laser, or who supplied the animals) and what happened (two minutes, perhaps). Take a minute or so to explain why your experiment or observation was reliable and repeatable – justify your experimental results, especially if it was an experimental design which requires people to think (a result-reversal design, perhaps). Then say 'So...' and tell them either 'You must change your minds about why there are so many sperm' or, much more often, 'You don't have to change your mind about this – A does indeed cause X!' Next a word about what you're going to do next is usually appropriate, and then *stop*, before or at 10 minutes.

Do not be afraid to do three unusual things in the question period: be prepared to wait for a little while if there's no instant question (your chairman should have one, though, if he's doing his job); do interrupt a questioner who's taking your question time simply to show that he's a clever guy – ask, 'What is your question, please?'; and do say, 'That's a question I haven't thought of – give me a moment!' if it is.

In a longer presentation the same points must be made, but of course it still might take only four minutes to give the audience the context; but now you have time to go more deeply into it. Well, perhaps, but that's usually a mistake! In the first part of your talk you can discuss the way you came to do the experimental design that you finally used; talk about the preliminary experiments, particularly ones that *didn't* give the expected result, and how you found out why. Make it a little climax when you finally come to getting the results: 'After two years we finally found three collisions which gave us the particle angles which we were looking for – and one had occurred within 10 minutes of the start of our project!'; ' ... and then the fish finally bred, on the last day of the project!' Use a couple of pictures. Do use anecdotes (brief and relevant) and cartoons if relevant and funny – (try them on colleagues first – but, remember, you are not a stand-up comedian). Tables of results with tens of little figures all over the spreadsheet are *not* good; use graphs, pie charts, simple-to-see-and-to-understand diagrams (see below, under poster presentations, for general instructions about diagrams, tables and visual information generally). Imagine it's your Aunt Mary, bright but not a scientist, in the audience when you're inventing the visual aids.

That was a little instruction – suggestions – for *how* to present your research work; much more difficult, requiring much thought, is *what* to present. While an informal research talk should not have the crystalline structure of a paper for publication (see below), a logical structure of much the same kind is usually best, most easily presented clearly and most easily understood. Particularly if you have an hour to talk, set up alternatives in the 'discussion of results' bit; then choose that alternative which gives most clarity, even if you don't give all the information that you would like them to have. Make sure that your audience can distinguish your negative, null and positive results. Talk about signal and noise (but perhaps not in those terms) and result-reversal theory, if you have used it. Importantly, remember to present your data, and not just the statistics you have generated from it, in graphical form. Keep the data *raw*, but not as tables with more than 100 numbers. Present the statistics as well, but only if they are both necessary and illuminating. And set up your discussion of *what it all means* so that your audience can see genuine alternatives (to A causes X, for example). Imagine your own prejudice would be disproved by interpretation no. 1, and generate others: no. 2, no. 3, etc. If you don't understand what we mean here, reread the first two chapters.

Other points which apply generally to public speaking (don't mumble, read your presentation, fiddle with your groin or pick your nose) we don't raise – but these are just as important. Make sure you get some specific feedback from your audience, especially in the early days. Pick someone who is an acquaintance – not a friend – and ask them to get opinions from others in the audience, and then report back to you – and arrange that you'll do the same for them. Not a friend, because people will know, and may be kind rather than honest! Realise the value of this advice[2].

However, remember that science is a series of stories, and that you are *telling a story* when you present your work: you have to hold your audience's attention. You always have a choice of endings, so choose the one that will change the most minds in your audience. You want them to think 'how interesting' and ask you lots of useful questions. Take time to go and listen to other speakers. Very few can *read* their story and hold the audience; practice to discover the style that works for you.

12.4 Saving theories

What happens if you have agreed to give an informal, or indeed formal, paper on your work and, two weeks (or two hours) before the appointed time *something goes wrong*? Let us imagine that your experiment has indeed worked but has given, unequivocally, the opposite result from what you expected. You have three alternatives: you can withdraw your paper, you can give the paper with some such remark as 'Although we haven't had time to properly assess these results, it looks likely that we shall have to change our minds about A and X' (the audience will understand from the split infinitive that you are upset!), or you can propose a *saving theory*.

Every result can be interpreted in the frame of the negative result with a saving theory. Simple saving theories are:

---

[2] Wear sunscreen – this you can trust us on!

- This is a statistical anomaly – we're repeating the experiment next week.

- The apparatus seems to have been contaminated with B (and here there is room to implicate colleagues, your mother-in-law or your supervisor).

- Although the technician told me that there was sodium chloride in the bottle as stated on the label, a preliminary test showed that....

- Previously we always did this experiment when the moon was full, but this time....

Saving theories which are less transparent, and which will get you the sympathy of some 5% of your critical colleagues, are:

- I had a car accident while carrying the last set of dog-mating samples, and it is possible that they got mis-identified.

- The samples were interspersed with another experimenter's in the radiation counter, and we're not sure what standard was being used.

- Or even – many of the mouse-egg culture samples, preserved in formalin, were drunk by a 4-year-old boy at my son's birthday party![3]

There are, of course, many more sophisticated saving-theory types:

- So it seems that cabbages growing on high phosphorus are much more attractive to cabbage-white butterflies, and that is why they are lighter in weight.

- Surprisingly, the rabbit whose serum we used had a myeloma, which produced this unexpected antibody to our test substances.

- Damn, we happened to take those neutrino readings at just the time when the neutrino flux from the supernova XXXX was passing through the Earth.

---

[3] These have all happened to JC, but none of the experiments were ruined, because in each case there were sufficient replicates to use instead.

These more sophisticated – and occasionally true – excuses for the wrong result may actually convince your colleagues too – even perhaps your external examiner!

One last point: we think that it's always worth telling your audience that *you* do know it's a saving theory. You get bonus marks for creativity.

12.5 Explaining and demonstrating

Almost always, your explanation of your own work within the context of the science will be a reductionist explanation: A causes X, provided that B is present; B causes X in the presence of an excess of A; cabbages need phosphorus. This is partly because you are adding another line to the stories, so it has to fit in. But remember that explanation does not necessarily provide understanding.

Refer to Section 10.2, where we considered how much credibility you have, and how much is needed. Some assertions we will believe without any requirement for demonstration; others we will not believe even with the most apparently convincing evidence before our eyes (the woman is *not* cut in half). We have already referred to the ease, but the inutility, of demonstrating that A causes X and the difficulty of convincing people of the converse. Further, if you have succeeded, by your brilliant experimental design, in showing that A does *not* cause X, you will be seeking grants from wholly different donors, at the very least.

In science, as in other social fields, your authority has some correlation with how clever you are, how often you have been seen to be right, and how good you are at convincing people. Too often, however, scientists who are right but have severe halitosis (we invent) find it very difficult to demonstrate that their theory is better than the orthodoxy. In such cases many of them will claim to be a misunderstood Galileo or Copernicus, and will spend increasing amounts of time and effort getting their ideas published. They begin by hoping for recognition by Royal Society journals, *Science or Nature*; then, because their cause is usually theoretical (real nature has very persuasive ways of dissuading practical scientists who are wrong...), they try *The Journal of Theoretical Biology* or *Speculations in Science and Technology*. In contrast, the practical scientist with a good experimental result, having used it to explain, can then use it as a demonstration. Many famous scientists have based their professional reputation on one of their experiments demonstrated in a lecture. It is not so popular (or possible, perhaps) now, but can be very convincing.

We have seen videos of such experimental results within PowerPoint presentations, which have the conversion power of a television evangelist – the expectation is that the audience will rise, chanting 'Yeah, yeah, I've seen the light'.

12.6 Poster sessions

If you have the choice – *don't* make a poster. They generally take more time to prepare than oral presentations (perhaps they should?), but always have less impact. The only advantage is that you have something to stick on the wall afterwards. But this is a personal prejudice (of GFM).

Nonetheless, this is a very common method of presenting scientific results, usually sets of results which are being considered for publication in refereed journals (and may have had several rejections). Most meetings of scientific societies, and meetings of scientific organisations focusing upon one aspect of a subject, will solicit posters from the participants. This is a complex political issue, because many of the participants can get their expenses paid by the home institution only if they can be a 'presence', by a talk or a poster, at the meeting.

It is, of course, in the presenters' interest to interact with colleagues, to be seen to be there and to be producing. Unfortunately, most of these presenters let themselves down badly. Even with today's desktop publishing technology, it is very common to see a majority of grotesquely inadequate posters, even – perhaps especially – at prestigious international meetings. Workers are tempted to write out scientific papers on their 1.5 m² boards, with tiny photographs, poor graphs and tables with hundreds of entries, and they all fall for the temptation. In the time set aside for the poster session, when they are to defend their assertions, one can imagine them saying 'See, here are all those results, two years' hard work; don't they justify my belief that high cadmium causes infertility?' They want to have all the data available, and they're right; but they want to have it all on display, and that's wrong.

The poster has the same problem as slides or transparencies , but much more acutely: how to give *enough* information to excite interest and, hopefully, understanding, but not overwhelm the eye of the beholder. Ideally, give just enough information to catch the eye.

Anything smaller than this (18 point) won't be read.

Similarly, your title for the poster should be readable from across the room, preferably in bright, contrasting colours. If your name will draw attention (or if you want to advertise it), have that up in lights too. If the subject is remarkable ('A New Position Adopted by Porcupines During Copulation'), make sure that it can intrigue people from afar. If you have a remarkable picture, make that a centrepiece. Some people are bringing projectors with PowerPoint displays or/and films that can be looped on a screen or a pale wall. We would recommend *not more than six* 'items' – text boxes, graphs, tables or pictures. Each should be readable from at least 2 metres away – 10-point type will enable you to write a lot, but no one will read it. On the other hand, don't overdo it. Ten telegraphic words in each text box may be too few, but 10 is usually better than 100! Perhaps 20–50 is about optimal on an A5- or A4-sized box, set out with some artistry so that the pattern is eye-catching.

Use colour; use any trick that will make it easier for your viewers to understand what you want to say. There is no way that a poster can be a presentation of your work – at best it should be an abstract, and the recommendations we make for the pre-abstract of a scientific paper below could serve as guidance to an economical poster presentation.

12.7 Giving a formal oral presentation

These usually occur later in your studentship, hopefully when you have discovered something of your own style in more informal presentations. Talking in front of a strange, hostile audience is not the time to experiment. Firstly, *know* what you are going to say; the key to a good presentation is the slides that you have prepared. A talk with good slides can be good or bad; a talk with poor slides is almost always bad. Don't be tempted by all the clever tricks you can do with PowerPoint; remember, it's *your* talk, not a PowerPoint demonstration! The slides themselves should follow much the same format as the suggestions we make for posters. What you include in the talk has much to do with the time available and the audience. It is good practice to be able to rewrite your presentations for different audiences, emphasising different aspects and putting a different 'spin' on the outcome. Your results on pig parasites will have differ-ent interpretations for the weekly meeting of the local Young Farmers Association, the British Parasitology Society annual meeting, and a research seminar at another university. This means that you have to be flexible in your understanding of the research you have done, that you are aware of the context within which the research is/can be viewed (see Chapter 5 again for this very important issue). Once you have decided on an 'angle', don't change halfway through. Especially, don't suddenly remember to describe the complicated statistics for the Young Farmers.

Otherwise the advice is much the same as for informal presentations: ensure that the talk is polished by presenting it to your colleagues within the research group. Get them to time and criticise the talk. Go to as many other such presentations as you can, and watch the process of presenting, so that you can pick up your own list of do's and don't's that match your own style. But don't worry unduly if you find that you can't *not* walk about, for example; make your problems part of your individual style. Be ready with your criticism for colleagues, and with comments on experimental design for people in the same field. And learn to receive these gracefully.

12.8 Writing a scientific paper

There are as many ways to do this as there are scientists, if not more (some of us are versatile!). This suggestion is for the beginner, who does not feel that s/he can claim the time of colleagues until there is some kind of a skeleton effort on the table to discuss and improve. The key is to remember that this, like a presentation, is a story, so decide *what the story is*. Like all good stories it should have a beginning, a middle and an end. And it should be an interesting story capable of changing minds. It should be clearly presented so that those who read it can criticise the methods, logic and the conclusions you draw, but *never* your results.

Introduction: Why the subject is interesting, where the hole is.

Materials and Methods: What was used, what I did with it, in enough detail that somebody else can do the same thing.

Results: What happened; what *actually* happened! (points on a graph are – or may be – *results*, but the *line* is discussion).

Discussion of Results: How repeatable, how reliable, do they address the question?

Theoretical Discussion: If the results are good, and like that, what does it mean? For example, is A really A? Sometimes it's a privative:[4] the *absence* of M! I saw X, but I wonder if it's ever been observed properly? It looked for a moment like Y, sideways on.

---

4 A privative is the absence of something: darkness, ignorance, death are privatives – their opposites (technically, complements) exist, but they themselves don't.

Conclusion: Have we filled the hole? What does the garden look like now?

To start is always the most difficult. Here are some suggestions:

- Pretend that you're six months into the future, and you find a copy of the paper. Imagine yourself picking it up. Now imagine what it said!

- Decide what figures/tables you want in the paper – draw them and write the paper from the middle out. This can be done only when you have sorted out the story in your mind already.

- Sit down with a clean sheet of paper, a sharp pencil and lots of quiet. (In our aged experience, even the friendliest computer screen seems less helpful – even for those two generations younger.) Write the first sentence, then the second and so on. Again, this can be done only when you have sorted out the story in your mind already.

- Faff around for ages, so that your collaborator(s) and/or supervisor(s) get bored waiting and write it themselves.

- Write the *pre-abstract* first: (if you haven't heard of a pre-abstract, that's because we've just invented it). The pre-abstract might have 15 or so sentences only, two or three each for Introduction, Materials and Methods, Results, Discussion of Results, Discussion and Conclusion. See the example in *Box 12.1*; you will probably find, as we did, that more sentences are needed in the Discussion sections, because there are links in the chain of argument. Note that this exercise is not an Abstract or a Summary; it is the skeleton of a paper, with enough of the soft tissue added that the overall shape can be seen. Once you've done that, turning it into the journal format should hold no terrors.

12.9 Writing a thesis

This is not easy. The best thing is to take advice from your supervisor and (more importantly) the other postgraduates and postdoctoral fellows around. If you have been writing up your research as publications, then you will have wisely kept all the paragraphs that got discarded, for inclusion in your thesis. Identify somebody to be the external examiner (it doesn't have to be *the* external examiner, but

Box 12.1
The pre-abstract

Introduction: Birds and mammals have coloured appendages, feathers and hairs; the colour is made in special cells, melanocytes, which donate pigment to the keratinising cells. Albinos have incompetent pigment-production, but what about white areas on coloured animals? Do they have no melanocytes at all, or melanocytes which don't make colour, or does pigment donation not occur?

Materials and Methods: Light Sussex chickens were used: they have black on wings and tail, but are white elsewhere; the chicks are totally white. Tissue explants from the base (collar) of growing feathers from white and black areas were treated in three ways: some were sectioned histologically, stained for pigment cells, and examined by various microscopic methods; others were implanted into the wing buds of White Leghorn 72-hour embryos, in the anticipation that any pigment cells present would colonise and pigment the chick's wing, or not; others were tissue cultured in a variety of natural-fluid media, in the hope that any pigment cell stock in the white feather explants would become pigmented in some media.

Results: No melanocytes were found by microscopy in white-feather explants, but plenty in those from black areas; these had long processes extending into the rows of epidermal cells that would become barbules. None of 3600 host White Leghorn eggs hatched, and no colonisation of embryo wings could be confirmed in the dead embryos. About half of the explants from white-feather areas showed abundant donating melanocytes in arterial blood-clot cultures, few in other media; the explants from black feather bases also failed to produce melanocytes in most media, even though good feather structure was formed, but they formed normal, pigmented feather in arterial blood-clot cultures.

Discussion of Results: The microscopic data on feather collar explants are consistent with the appearance of the feathers in those areas; donation could be seen. The White Leghorn transplant attempts gave a null result. The arterial blood plasma clot culture medium gave good control black feather from black feather collar tissue, but also provoked black melanocyte formation in cultures from white feathers. That only half of cultures gave this result does not reduce the reliability of the result. It was repeated in four culture series, from six birds. This is a firm positive result: there are pigment cells in the white areas, which can be provoked into visibility by their making melanocytes.

Theoretical Discussion: Microscopic data was not helpful, because the melanocytes of the white areas were non-existent or non-pigmented; had they been 'continent', making but not donating melanin (as in frog skin), this would have been seen. The null result from the Rawles technique is to be regretted but does not affect the issues. Had just a few of the white-feather-collar explants produced melanocytes, this would have been very strong evidence for the presence of pigment cell stock; in the best medium, however, about half did. So we may be confident that there is competent pigment cell stock in the white feather collars, but it is not being 'turned on' to make pigment. Whether there are uncoloured melanocytes, perhaps donating unpigmented granules, is not certain; the microscopy would probably have brought them into view and did not, so there is no reason to postulate them. On the other hand, black feather explants failed to make pigmented melanocytes in most of their cultures, further evidence of the sensitivity of the melanocyte system to outside influences. This brings into consideration birds like the Brown Leghorn, or pheasants, with many colours from the same genotype of pigment cells: perhaps the different colours are also the reflection of different milieux for the pigment cells.

Conclusions: White areas of coloured-and-white animals can have competent, but not effective, pigment cells. Perhaps there are many kinds of cells in all tissues that have not produced their functional form but form a reservoir of 'stem cells'.

somebody in the field who *might* be the external examiner) and write it *for* that person. This will help you keep the detail at the right level and not to go on forever about the basic biology of the porpoise or how cyclotrons work (because the external examiner knows that already). You might also develop the art of self-criticism, which is perhaps not self-criticism at all, because you usually have to pretend to be somebody else to do it.

There's one kind of very general mistake that postgraduates all seem to make. JC has called it 'the temperature was measured' mistake, and you can find it in the abstracts of about a third of published papers. What you should say, of course, is that 'The temperature was 37 °C' – not being mysterious and saying only that you measured it. This translates into a more general point about scientific writing, and especially a thesis. One of GFM's colleagues puts it very elegantly. The science leading up to the thesis is like a Sherlock Holmes story: it's all mysteries, blind alleys and red herrings, but in the end you find out that the butler did it. But when you write it up, you should start with 'The butler did it' and then explain how you know. Don't be tempted to try and lead the reader through an autobiographical account of your trials and tribulations and describe all the (false) clues you found on the way. You are presenting the evidence, so should only include the relevant facts that contribute to the case. Hopefully, students learn this – then when they become supervisors they can pass it on. This is how the format of a scientific Ph.D. thesis has evolved. But bad writing is only a symptom. Most people don't say that the temperature was measured – they can distinguish the important detail (i.e. what the temperature was) from the trivial information (i.e. that the temperature was measured).This is how the culture of science improves.

Otherwise, the only advice is to do it – glue your bum to the seat and don't get distracted into making another coffee, hoovering, reading the book that you have *always* wanted to read but never got round to, taking a trip to India to discover your soul, having a baby, waiting for somebody else's comments, criticisms or results, waiting until you have *that* reference, etc. If postgraduates could do science like they can make excuses for not writing up…

Answers for Mind Games on page 27

Answer 1: Tin hats protect heads, but they are not invincible. The numbers of soldiers dying from head injuries in the trenches decreased. The increased number of injuries are those soldiers whose lives had been saved – without the hats they would have died. Now remember what it was like not to understand.

Answer 2: This is a trick because you have to separate each square from every other square, that is, there is no 'clever' answer. The trick is that there is no trick. So, no matter what order you do them, you have to break along each of the $3 \times 7 = 21$ lines. Yes or no?

Postamble

We have a last thought for you postgraduates, hoping to get hints from us about the important thoughts which you have to invent, to make your M.Sc. dissertation really masterful, or your Ph.D. thesis really philosophical. This thesis is, after all, the major product of years of work.

No, it isn't.

If it is, you have failed in your learning to be a scientist, and we don't want you in our club. If you genuinely believe that your thesis, documenting your discovery about pigment cells or protons or politeness is the most important product of the time you have spent engaged in scientific activity, forcing the universe to reveal her secrets, we hope you are wrong. We would like to think that this piece of work, however beautiful and mind-changing (or not), is simply a witness to a change in *you*. You will understand, as we did after our *viva voce* examinations, that *your* development of *your* abilities to engage with problems scientifically has given you the right to self-election to scientist. We hope that you will never believe things just because you want to.

Welcome to a life whose rewards are mostly not monetary; whose central paradox is that you put most effort into attempted disproof of the things you most want to believe; whose demonstration of close and valued friendship is honest, perhaps impolite, criticism. You will learn to live with the certain knowledge that your valued fundamental concepts – which led to your understanding of how the universe works, and which you received from those whose respect you have finally earned – will look silly in 50 years' time, just as the ideas of 50 years ago look now. You will learn that very few of us are Newtons, Darwins or Haldanes, and that being such requires some luck and having the courage to seize opportunities when they come.

Welcome to one of the most honest and human of human activities.

Index